THE PULSE OF THE BAY 2022

1972

2022

50 YEARS AFTER THE CLEAN WATER ACT

A REPORT OF THE REGIONAL MONITORING PROGRAM
FOR WATER QUALITY IN SAN FRANCISCO BAY

SUGGESTED CITATION: SFEI. 2022. The Pulse of the Bay: 50 Years After the Clean Water Act. SFEI Contribution #1095. San Francisco Estuary Institute, Richmond, CA.

OVERVIEW

Jay Davis, San Francisco Estuary Institute

The Bay Transformed

This August an algal bloom of historic proportions blanketed San Francisco Bay. The most concerning aspect of the bloom was the death of thousands of fish across wide areas of the Bay, including 10,000 yellowfin goby, hundreds of striped bass and white sturgeon, and other species like green sturgeon, bat rays, and sharks. This is the most widespread and deadliest fish kill in the Bay in recent memory.

Back in the late 1960s and early 1970s, however, as documented in a 1972 US Environmental Protection Agency (USEPA) report, die-offs of thousands of fish were reported on nearly an annual basis, with a peak of over 100,000 fish (including over 90,000 striped bass) in 1965. The listed causes of these fish kills included oil and refinery waste, sewage, low dissolved oxygen, high salt, and algal blooms. These fish kills were emblematic of an era when the Bay and its shoreline were used as a dumping ground for minimally treated sewage, industrial wastewater, polluted runoff, and solid waste. As a result, Bay water quality was abysmal: unsuitable for aquatic life, unsafe for swimming or consuming fish and shellfish, and as a 1941 report on sewage disposal commissioned by seven East Bay cities had put it, "obnoxiously and notoriously foul and an affront to civic pride and common decency."

This year marks the 50th anniversary of a transformational turning point for Bay water quality: passage of the Federal Water Pollution Control Act Amendments of 1972, commonly known as the Clean Water Act (CWA). The CWA provided a legal framework and a considerable amount of federal funding (over $1 billion, equivalent to approximately $7 billion in 2022 dollars) that drove a rapid and remarkable improvement in Bay water quality. By 1987, all municipal wastewater treatment plants but one were providing secondary treatment, which effectively removes oxygen-demanding organic matter and bacteria as well as many toxic metals and organic chemicals. Bay monitoring data available for the 1970s and 1980s show that dissolved oxygen levels increased, and bacteria and toxic metal concentrations sharply declined. By 1982, public harvesting of shellfish in San Mateo County was approved for the first time in 50 years. By 1987, the Water Board concluded that swimming was safe in most areas of the Bay during summer.

On the occasion of this momentous milestone, Section 1 of this edition of The Pulse of the Bay includes nine perspectives written by representatives of the groups that have a prominent role in managing Bay water quality, including state and federal agencies, municipalities, industry, dredgers, and the leading environmental group advocating for Bay water quality. Section 2 includes profiles of the parameters that have been the main water quality concerns over the past 50 years, with a focus on long-term trends and a historical perspective.

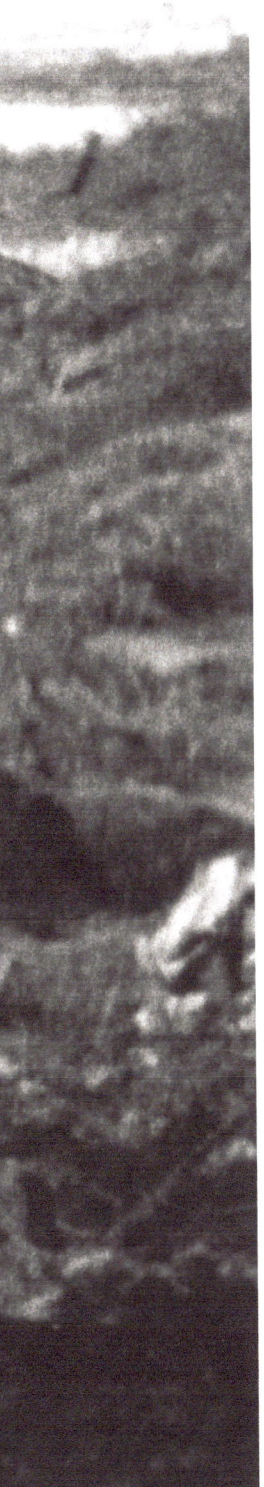

◀ **A tributary to the Bay in 1965.** This image and several others in this report are from "No Deposit, No Return," an outstanding KRON-TV Assignment Four special report documenting the impact of waste disposal on the Bay. Used with permission from KRON and assistance from the Bay Area Television Archive, a collection maintained by the Leonard Library of San Francisco State University. *Imagery courtesy of Bay Area TV Archive at SF State University.*

Water Quality Successes Under the CWA

In addition to driving those rapid early improvements resulting from improved wastewater treatment, the CWA also provided a legal and regulatory structure that has been used to achieve many other successes for Bay water quality. Most of the noteworthy wins have resulted from creative, innovative, and collaborative implementation by the Bay Area water quality management community of the elements included in the Act and in subsequent amendments. The following list notes some of the successes described in the nine perspectives, all of which relate to provisions of the CWA.

Collaborative Monitoring • The San Francisco Bay Regional Water Board (Water Board) made creative and innovative use of its CWA permitting authority to encourage dischargers to initiate cooperative regional monitoring in 1993, replacing the uncoordinated monitoring of individual discharges that had occurred previously. The Regional Monitoring Program for Water Quality in San Francisco Bay (RMP) quickly made the Bay one of the best-monitored waterbodies in the world. The RMP is governed collaboratively by regulators and permit holders, and this has fostered an atmosphere of trust and joint fact-finding that has minimized conflict in the regulatory process. The collaborative RMP model was followed by the Nutrient Management Strategy, initiated in 2012 to provide the science to inform decisions regarding the expensive measures that would be needed to reduce nutrient inputs to the Bay.

Collaborative Management • The foremost example is the Dredged Material Management Office (DMMO), which was created in 1995 to provide more effective and efficient regulation of proposed dredging and dredged material disposal activities. The DMMO agencies include the US Army Corps of Engineers, the Bay Conservation and Development Commission, the Water Board, the State Lands Commission, and the USEPA.

Innovative Regulation • The Bay Area has been a national leader in urban stormwater management. The stormwater discharge permit issued to municipalities in Santa Clara County in 1990 was the first such permit issued in the US with pollutant control requirements. In 2009 the Water Board made another major stride forward in the management of urban stormwater through the adoption of a new Municipal Regional Stormwater Permit (MRP) that covers much of the Bay Area.

Proactive Monitoring • The Nutrient Management Strategy was initiated in 2012 to provide the science needed to help prevent problems related to nutrient enrichment before they occur. The RMP is a world leader in monitoring contaminants of emerging concern in the aquatic environment, with the goal of early identification and early management intervention.

Creative Solutions • Initiated in the Bay Area, the Brake Pad Partnership was a collaboration of stormwater agencies, brake pad manufacturers, water quality regulators, and environmental groups whose work led to the passage of state legislation requiring that the amount of copper in brake pads sold in California be reduced to no more than 0.5% by 2025. Regional collaboration and innovation are also yielding multi-benefit, nature-based projects like the Oro Loma horizontal levee, which protects the shoreline against the adverse effects of sea level rise while removing nutrients and other contaminants from wastewater and providing habitat.

Funding • In 2008, USEPA began the San Francisco Bay Water Quality Improvement Fund to accelerate wetlands and water quality restoration. To date, more than $71.4 million in grants have been awarded to 80 different partners, and another $29 million will be awarded in 2022 alone.

Contaminant Successes • Improved science and management actions have led to several contaminants being removed

◀ **Radhika Fox, USEPA Assistant Administrator for Water.** Speaking at an event on September 16, 2022, with the Bay in the background, to celebrate the 50th anniversary of the Clean Water Act. *Photograph by Shira Bezalel.*

OVERVIEW **V**

from the CWA Section 303(d) List of impairments of Bay water quality, including copper, nickel, and diazinon. RMP monitoring thoroughly documented the successful reduction of polybrominated diphenyl ether (PBDE) flame retardants in the Bay as a result of a California law banning their use.

Contaminant Control Plans ▪ Carefully considered control plans have been crafted to address other, more challenging contaminants that remain on the 303(d) List. Significant efforts have been made to reduce loads as well, such as remediation of mercury mining waste in the Guadalupe River watershed, reduction of selenium loads from oil refineries, and identification and abatement of watershed source areas for PCBs.

Challenges Ahead

While there are many CWA successes to be celebrated and much progress has been made, the damage being wrought by the current Bay-wide algal bloom is a grim reminder that we have not fully met the goals set in the CWA or by the regulatory processes it established. The CWA set ambitious goals of all waters of the US being fishable and swimmable by 1983, and eliminating all pollutant discharges by 1985. Decades later, however, the Bay is only partially fishable, not fully swimmable, and not fully safe for aquatic life. Many contaminants have remained on the 303(d) List since the 1990s, and some of those of greatest concern won't be de-listed anytime soon. The Bay continues to face a number of ongoing and new water quality challenges, as described in the perspectives in Section 1 and the parameter profiles in Section 2. A few of them are briefly noted here.

> **Infrastructure** ▪ Much of the infrastructure for managing municipal wastewater and stormwater is now 40 years old or more and in need of upgrades. This is in addition to the need for new infrastructure to improve removal of nutrients, legacy contaminants like PCBs and mercury, and emerging contaminants.

Nutrients ▪ While the cause of the current bloom is not yet fully understood, its occurrence is not a major surprise given the longstanding recognition and concern about the Bay's high nutrient loads and concentrations.

Contaminated Sites ▪ While cleanups of some source areas in the watersheds and contaminated sites in the Bay have been conducted, others areas and sites have not, and in some cases the cleanups have not been fully effective.

Stormwater Loads ▪ Stormwater mostly flows to the Bay untreated and is a major pathway for legacy contaminants, emerging contaminants, and bacteria. Reducing these loads has proven to be challenging.

Emerging Contaminants ▪ The number of chemicals in commercial use has passed 100,000 and continues to grow. Identifying the small subset of these that pose significant threats to Bay water quality is a formidable challenge, as is managing the ones that have already been identified (like PFAS, toxic tire additives, and microplastics).

Climate Change ▪ More severe droughts can lead to higher concentrations of contaminants like selenium in the North Bay. On the other hand, more intense storms can potentially mobilize larger loads of contaminants from Bay watersheds.

Water Conservation ▪ Increasing demand for drinking water relative to supply is driving a move toward more recycling of municipal wastewater and stormwater, with concerns for funding the needed infrastructure and managing the waste from reverse osmosis treatment projects.

A Bay warning sign in 1965 from the "No Deposit, No Return" documentary (*courtesy of KRON-TV and Bay Area TV Archive at SF State University*).

Health notice for swimmers in 2022 (*photograph by Shira Bezalel*).

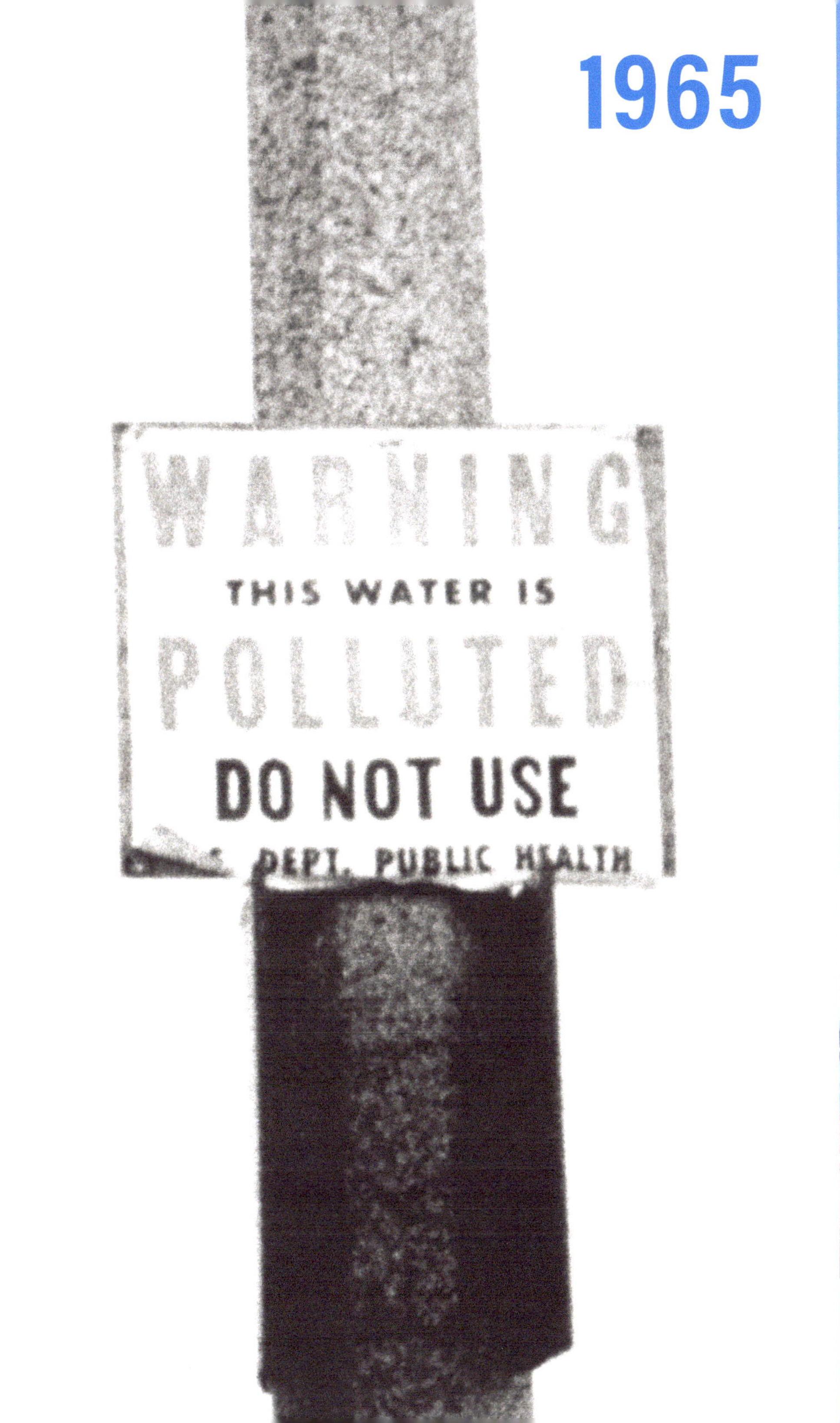
WARNING
THIS WATER IS
POLLUTED
DO NOT USE
DEPT. PUBLIC HEALTH

1965

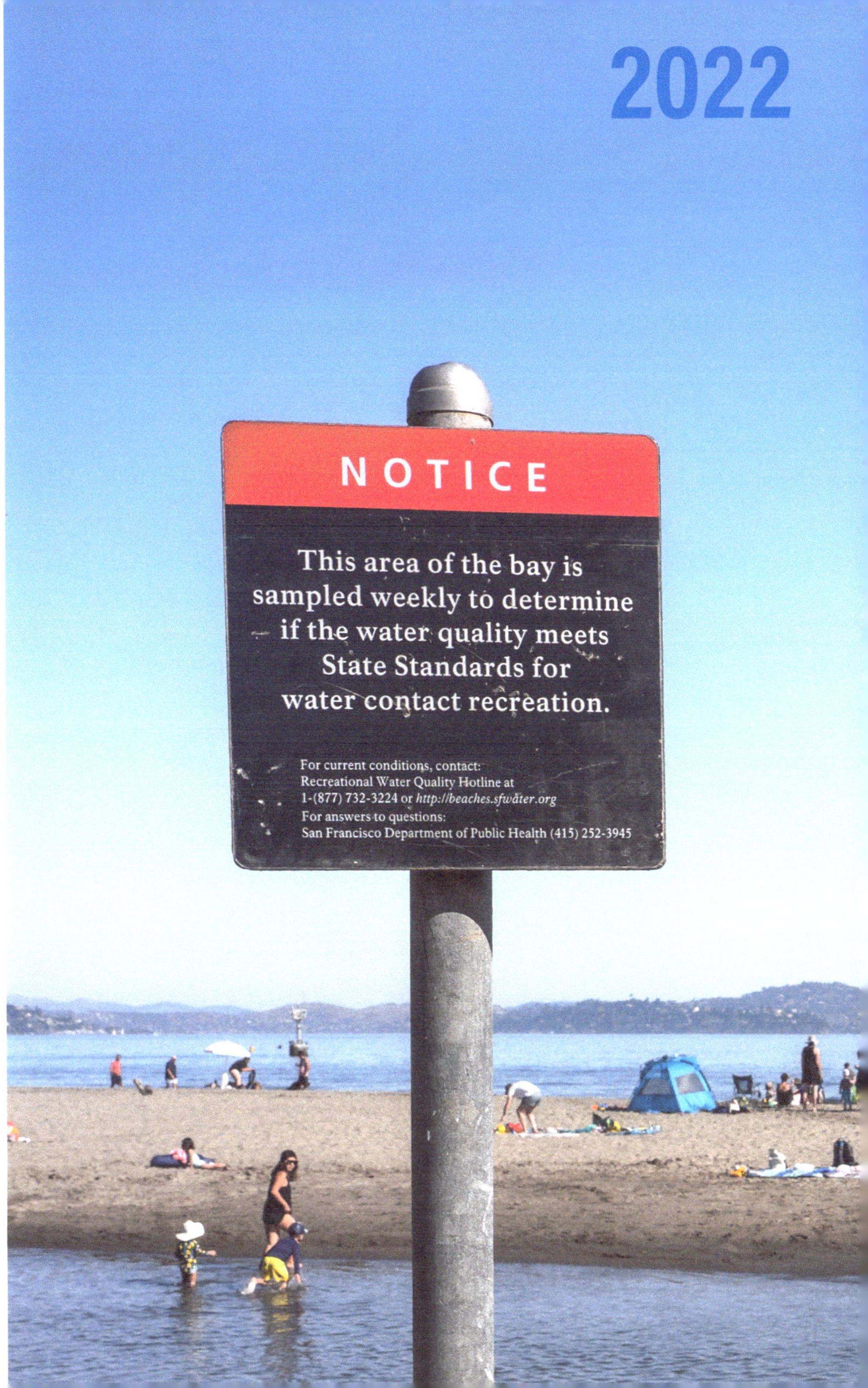
NOTICE
This area of the bay is sampled weekly to determine if the water quality meets State Standards for water contact recreation.
For current conditions, contact:
Recreational Water Quality Hotline at
1-(877) 732-3224 or http://beaches.sfwater.org
For answers to questions:
San Francisco Department of Public Health (415) 252-3945

2022

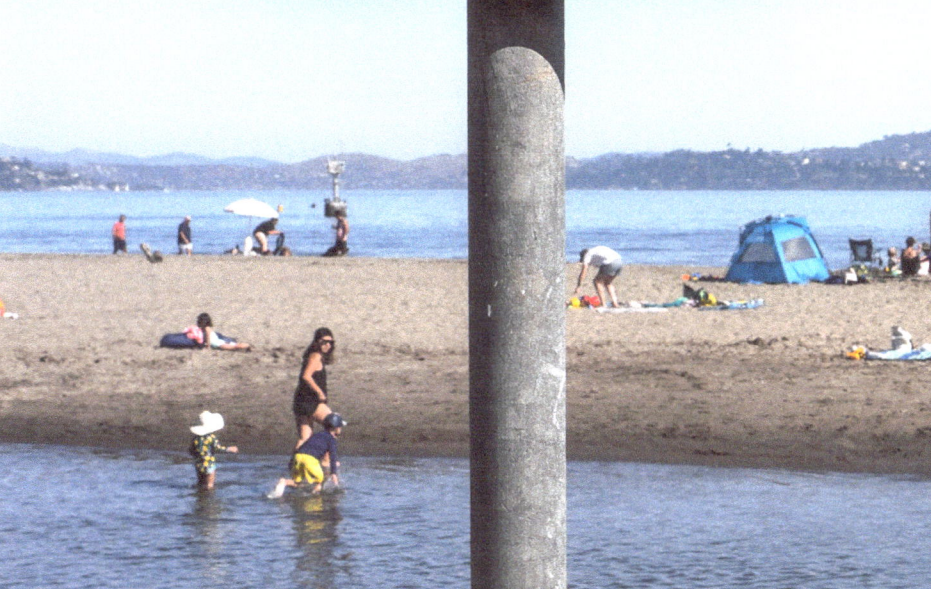

Crissy Field, ▶ September 2022 (*photograph by Shira Bezalel*).

Sea Level Rise ▪ Concerns for protecting the Bay shoreline and low-lying property have heightened the need for sediment management, including the reuse of dredged material. Another concern is the potential for remobilizing legacy contaminants by inundating contaminated soils or by rising groundwater.

Environmental Justice ▪ Many of the impacts of these issues and others are most acute in underserved shoreline communities that depend the most on a clean Bay for subsistence and their physical and mental health.

The need for collaboration, innovation, and investment is as great as ever if we are to succeed in rising to these challenges and meeting the water quality goals for the Bay set under the Clean Water Act. The RMP will continue to play a crucial role in providing the scientific foundation for managing Bay water quality. §

CONTENTS

Jay Davis • San Francisco Estuary Institute • **OVERVIEW**   ii-viii

PERSPECTIVES   1-45

Alexis Strauss-Hacker	**2-5**
Unsafe for Swimming (1938)	**6-7**
Jim McGrath	**8-9**
Strawberry Creek (1941 and 2022)	**10-11**
San Francisco Bay Regional Water Quality Control Board	**12-19**
Marin Headlands (1965 and 2022)	**14-15**
Aquatic Park (1965 and 2022)	**20-21**
US Environmental Protection Agency	**22-23**
Union Sanitary District (1972 and 2022)	**24-25**
Bay Area Clean Water Agencies	**26-27**
The Evolution of the Bay Area's Largest Wastewater Treatment Plant	**28-29**
Bay Area Municipal Stormwater Agencies	**30-35**
Dredgers	**36-39**
US Army Corps of Engineers	**40-41**
Remnants of the Dumping Ground Era	**42-43**
Baykeeper	**44-45**

WATER QUALITY PARAMETER SUMMARIES   46-73

Bacteria	**48-49**	Selenium	**62-63**
Organic Waste	**50-51**	Copper	**64-65**
Nutrients	**52-55**	PBDEs	**66-67**
Mercury	**56-57**	PFAS	**68-69**
PCBs	**58-59**	Bisphenols and OPEs	**70-71**
Dioxins	**60-61**	Microplastics	**72-73**

303(d) List	**74**
Status of Pollutants of Concern	**74**
RMP Committee Members and Participants	**75**
References	**76**
Credits and Acknowledgements	**77**

Many of the key resources cited in this report, including the KRON documentary "No Deposit, No Return" and the USEPA Documerica photo collection, are available via <u>the web page for this Pulse</u>

PERSPECTIVES

1968

50 YEARS AFTER THE CLEAN WATER ACT

2022

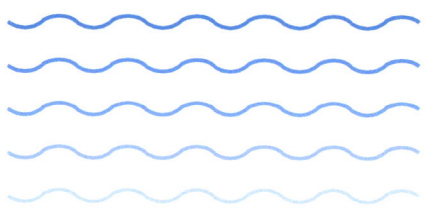

Alexis Strauss-Hacker

*Vice-Chair,
San Francisco Bay Regional
Water Quality Control Board*

We've been well-served by 50 years of the Clean Water Act in addressing major public health and environmental challenges, in building wastewater and stormwater infrastructure and controls, and protecting the beneficial uses of water. As we've faced new challenges, such as new classes of pollutant, some have sought federal legislation as the primary regulatory tool. We've weaned ourselves from this approach, of necessity. Throughout California, and specifically in the San Francisco Bay Area, many of our water quality, water supply, habitat, infrastructure, and species problems are being met with collaborative data-gathering, evidence-based analysis, robust funding, innovation, and risk-taking. We can't readily produce numeric or narrative criteria in each State (or authorized Tribe) to tackle the vast array of emerging contaminants, but we can be effective in other ways. For example, in just a few years, the Water Boards have begun closely tracking data from PFAS-affected sites, and scientists, wastewater systems, the Centers for Disease Control and Prevention, California Department of Public Health, and State Water Resources Control Board are analyzing influent wastewater for SARS-CoV-2 pandemic trends. The Bay Area's 37 POTWs are tackling nutrient pollution with the Regional Board and the San Francisco Estuary Institute, investing in the science in advance of developing optimal controls.

In 2007, observing the 35th anniversary of the Clean Water Act in the Pulse, I noted: "Now, after 35 years of our collective experience implementing federal and state water pollution programs, our focus is both on maintaining the gains achieved, and addressing a worthy set of new challenges." This is still valid, though our challenges are yet more complex, testing our ability to work together effectively and make wise decisions with the data and funding available to us.

Infrastructure

In recent years, multi-benefit projects are reaping rewards for those who diligently pursued the concept, funding, and permits needed. The Oro Loma horizontal levee project is one noteworthy example, with its proven ability to remove key contaminants while providing habitat. We want to encourage more innovation. With available infrastructure funds, we also need to spur more wastewater reclamation and reuse. The current level of Bay Area water reclamation is less than half that of southern California. While we depend on the initiative coming from each wastewater utility to reclaim and reuse water with its customers, what can and should the Regional Water Quality Control Board do to incentivize greater investment in this area, to build now for a better long-term outcome? As we're all acutely aware of the need to more efficiently use drinking water, stormwater, and wastewater, we need more low-impact development, more on-land containment/infiltration/reuse of stormwater, and more wastewater reclamation/reuse to sustain our Bay Area population to align better with the water supply available to us.

Underserved Communities

It is 2022, and yet we still have Bay-adjacent communities lacking stormwater infrastructure and drainage, communities which flood, where pollution is mobilized and eventually reaches the Bay. I'm hopeful we can encourage these cities/systems to take advantage of the USEPA/State Water Board infrastructure funds to make needed investments in the next couple of years. In some communities, access to the Bay for recreation and subsistence fishing needs to improve, facilities with recurring odor and spill complaints need a more coordinated response from those of us in government, and idling truck traffic brings significant pollution to certain neighborhoods. Working with CalEPA's environmental justice program, the Water Board, Bay Area Air District, USEPA and others can work together to hear and respond to communities' key concerns. We are expanding our ability to reach communities in the language(s) spoken, and in particular, we want to do more to engage students in understanding their concerns and our responsive efforts.

Dredging

As other contributors to this issue will address climate change, shoreline resilience, and nature-based solutions, I'm confident we can work effectively with the US Army Corps of Engineers and local dredging project sponsors to bring more dredged material into Bay Area locations for beneficial reuse. We've long been focusing on how to more economically move dredged material to where it can augment levees, nourish beaches, and support other suitable applications, and I'm optimistic this year may bring some welcome changes.

Contaminated Sites

The Regional Water Quality Control Board, the Bay Conservation and Development Commission, and the California Department of Toxic Substances Control are focusing on contaminated sites at the edge of the Bay, where pollution may be mobilized by wave and tide action and changing sea levels. We want to bring greater urgency to cleaning up these sites and will engage with the communities which host these sites. While we have used mechanisms such as Total Maximum Daily Load implementation plans for mercury and PCBs, we have not yet realized the needed gains in reducing PCB loadings to the Bay and fish populations, and thus a more concerted emphasis on the most contaminated industrial sites at the Bay's margin is now needed.

Vulnerable Rural Systems

Beyond the Bay Area's urban setting, we have pressing needs in rural areas for upgraded drinking water and wastewater systems. In California alone we have several thousand small drinking water systems, many struggling to serve safe and affordable water. The many obstacles to technical, financial and managerial operation of small water systems are a function of scale – a small system may have one part-time or volunteer operator and lack the fee revenue to make needed operational upgrades. We can use available drinking water infrastructure funds to increase technical assistance to small systems (such as through circuit riders and other person-to-person means), but can we wean ourselves from the unsustainable but staunchly defended approach to "local control" of several

> *Many of our water quality, water supply, habitat, infrastructure, and species problems are being met with collaborative data-gathering, evidence-based analysis, robust funding, innovation, and risk-taking*

thousand small systems? While individual consolidation of a few systems has been achieved, we still have dozens of systems which today face imminent lack of water to serve or lack safe water to serve. We will spend much time and money on meeting these immediate needs and have yet to forge consensus on a more systematic, statewide solution.

Rivers and Oceans

Even with very positive momentum, significant threats to water quality and biota, such as posed by microplastics and marine debris, still lack effective large-scale solutions. The vast expanse of ocean seems beyond our comfort zone, even as state legislative actions and a robust growth in microplastic scientific analysis continue. Moreover, we have no nimble tool to deter or respond to vessel abandonment and its associated pollution as boats founder and sink, whether on the Pacific Coast, in the Oakland Estuary, or in the Sacramento/San Joaquin Delta. We continue to measure critical declines in fish populations, even as we strengthen fish-consumption protections for more vulnerable consumers (e.g., mercury water quality criteria) and adopt tribal beneficial uses into Basin Plans.

> *I'm encouraged and inspired by the many dedicated and accomplished individuals working to address Bay Area water-quality challenges, who lead organizations and develop the next generation workforce with their outstanding examples and collaboration*

Workforce and Leadership

I'm encouraged and inspired by the many dedicated and accomplished individuals working to address Bay Area water-quality challenges, who lead organizations and develop the next generation workforce with their outstanding examples and collaboration. We can do more to showcase our talented staff and encourage the ability to serve in different local, regional, state, and federal roles, even temporarily, to broaden our understanding and perspectives. I welcome your comments, corrections, and ideas at Alexis.Hacker@waterboards.ca.gov §

Fishers at Piers 30-32 in San Francisco ▶
(photograph by Kelly Moran).

Photograph courtesy of the San Francisco History Center, San Francisco Public Library

1938

Unsafe for Swimming

The Gilman Beach Playground was constructed just north of Candlestick Point in the early 1930s by the Works Progress Administration as part of the New Deal. The accompanying description (to the right) is an indication of how the discharge of untreated sewage prior to the CWA made the Bay unsafe for swimming, and that there was monitoring and awareness of the problem.

This Beach Playground was not there for long — in the 1950s this area of the Bay was filled in as part of the development of Candlestick Park, which opened in 1960. Gilman Playground, without a beach, is still present at the same location.

"The water at Gilman Beach Playground has been found to be polluted by health authorities, hence swimming there is frowned upon by the Recreation Commission, which directs it. But it's hard to keep youngsters out of the water (witness scene pictured here) and city officials are looking forward to the day when a sewage project will correct the condition and make the waters safe for swimming.

Until then, the Recreation Commission is concentrating its efforts on the boating and maritime phases of the unique project. Eventually, they believe, the cove pictured in upper center photo will be dotted with small craft of every description; there will be projects to teach young visitors all types of sea lore, the etiquette of deck games, the construction of model boats and kindred subjects."
— Newscopy from 1938

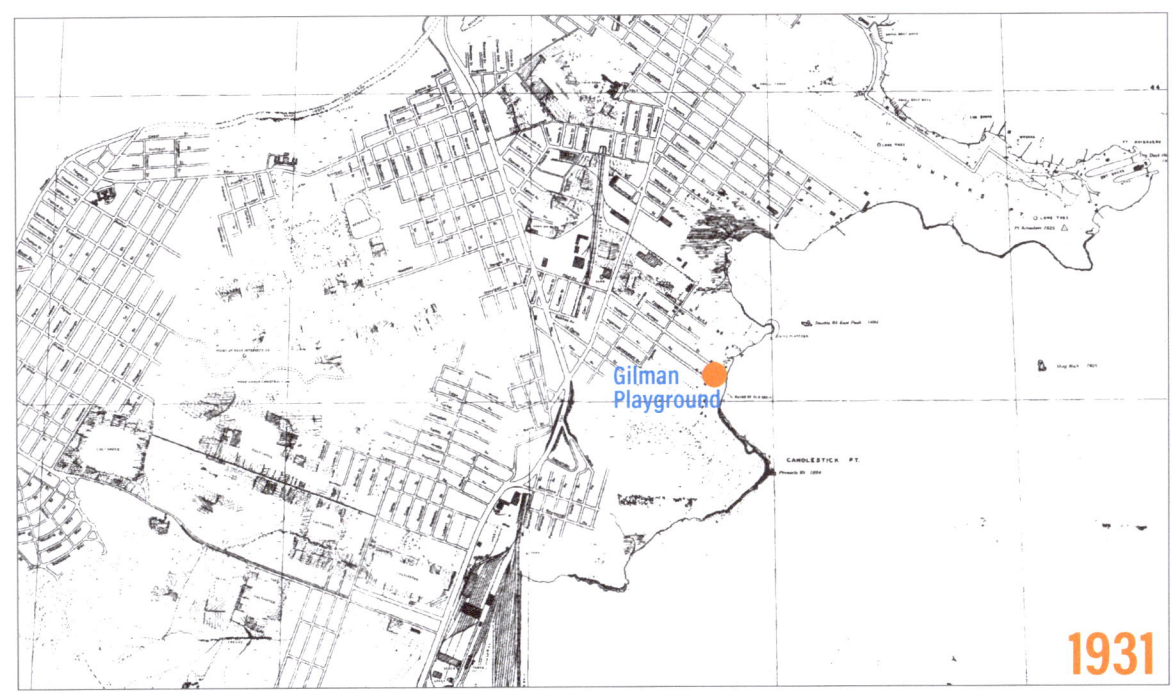

1931

Above · Map courtesy of NOAA, U.S. Coast and Geodetic Survey T-sheet, 1931
Below · Imagery courtesy of Google Earth, 2022

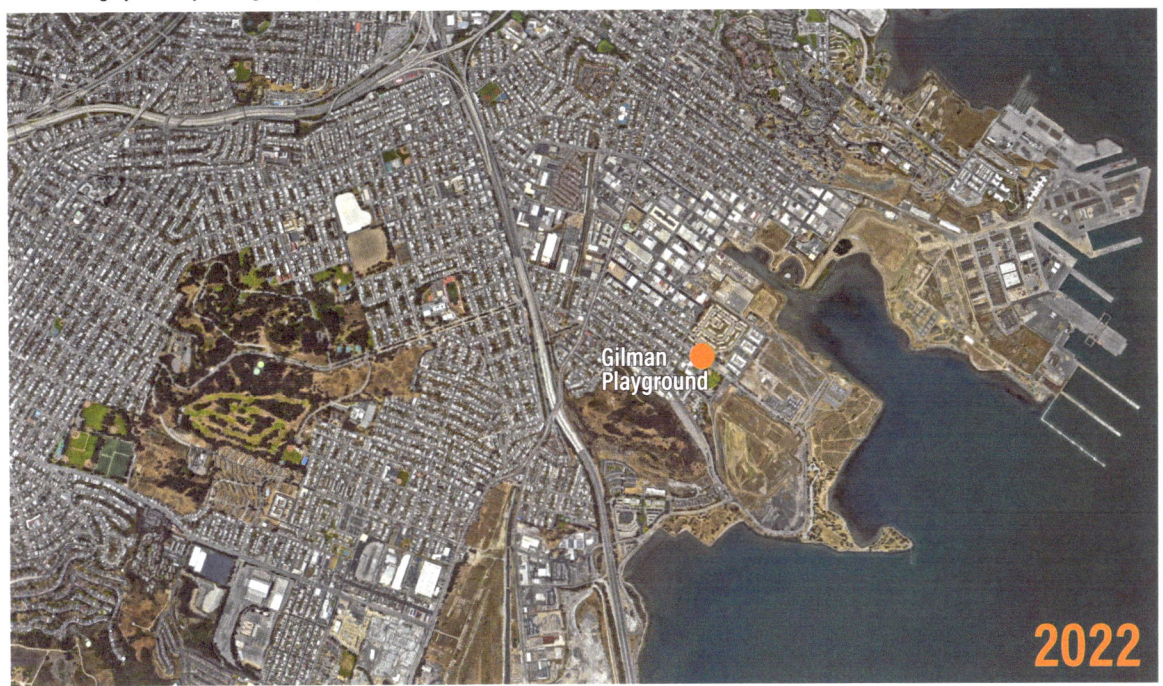

2022

PERSPECTIVES: 50 YEARS AFTER THE CLEAN WATER ACT **7**

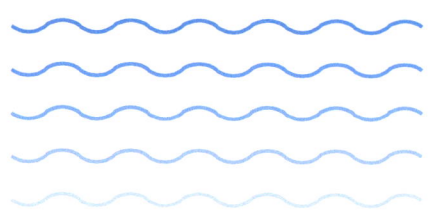

Jim McGrath

Member of the San Francisco Bay Regional Water Quality Control Board from 2008-2022

I **started my career as a work study student at the US Environmental Protection Agency (USEPA)** just after the Clean Water Act passed in 1972. It is gratifying to look at the progress that has been made, particularly in California. It is important to recognize the partnerships that account for that progress, and renew and broaden the partnerships that will be required to deal with the difficult issues that remain. I'll not repeat the wise comments by my colleague Alexis Hacker, but try to review some of the successes and future challenges that we face.

Successes

Clean up of point sources and protection and restoration of wetlands and streams have to be celebrated. I remember my shock when I walked into some of the urban sewage treatment plants while I was working on the construction grant program for USEPA. Now we have publicly owned systems that outperform their regulatory requirements, with professional staff proud of their work and partnering with us in addressing ways to reduce nutrient loads. Similar progress has been made with most industrial facilities, and the dramatic reductions in loadings have been well reported in previous Pulse articles.

I spent years struggling to protect wetlands, and I am eternally thankful for the partnerships with USEPA and the Corps of Engineers that established rigorous protections and mitigation requirements that corrected some of the early failures. After working for years to prevent a publicly funded marina from being constructed in the Bolsa Chica wetlands in Huntington Beach, I had the good fortune to tour the restored site with three of the partners that helped create a better system. Similar successes are underway with restoration of salt ponds in both the North and South Bays, thanks to the support and vision of Senator Diane Feinstein and the commitment and negotiating skills of Will Travis while he was executive director of the Bay Conservation and Development Commission that helped achieve public ownership of most of the wetlands along San Pablo Bay. Just weeks ago, we celebrated the establishment of a second fish ladder on Alameda Creek in Fremont—one of our best hopes for restoring salmonid runs along the Bay. I was fortunate to work for the Port of Oakland from 1990 until 2006, when protecting and restoring wetlands was in the Port's interest. I worked on restoration projects along San Leandro Bay, at Sonoma Baylands, Hamilton Airfield, Montezuma Wetlands, and Middle Harbor in Oakland—the latter four using dredged material and federal funding for restoring hundreds of acres. All of these efforts required citizen activists and public funding from organizations like the Coastal Conservancy and the US Fish and Wildlife Service.

While the Bay is not yet fishable in the sense that all fish are safe to eat, we have programs in place to reduce the historic loads of mercury and PCBs that are the greatest threat, and educational programs to make sure that people who catch fish consume them with minimal risk. Swimming has become common in the Bay, particularly in the parts of the East Bay which I visit. Swimmers, and indeed all who recreate on and near the Bay, are important stewards and advocates to organize to support protection and restoration of the Bay—a position that you might expect from someone who has windsurfed on the Bay for more than 40 years.

Renewing our Storm Water Systems

While the early efforts of USEPA and the Water Boards properly focused on point sources, those of us working on water quality knew that we would have to deal with urban runoff and contaminated sites to achieve the fishable and swimmable standards. We didn't realize in the earliest days how difficult it would be, both politically and technically. Not long after I was appointed to the Regional Board—more than 30 years after the Clean Water Act was passed—I realized that cleaning up the urban slobber that makes its way to the Bay would require a far more ambitious collaboration than I had thought. Fortunately, the aging and deterioration of the stormwater systems in most of the older cities had also convinced many local governments that they had to renew those systems with a greener approach that might be easier to maintain. Climate change, and the certainty of rising sea levels and groundwater added conviction and at least some sense of urgency to this effort. Fortunately, there has been an unprecedented amount of money brought to bear, between Measure AA, fostered by Save the Bay, and the investments in the Americans Recovery Act and 2021 state legislation. With at least some funding available, we need to turn our attention to ensuring that we are addressing the highest priority threats of water quality and flooding, and we are nurturing the public support that will allow us to spend the next few decades rebuilding those systems to deliver cleaner water and a resilient landscape.

Challenges

Our legacy of ignoring environmental impacts of development have left many contaminated sites, predominantly in disadvantaged neighborhoods. Often, clean-ups are difficult because contaminants are bound with soil. Improving our environmental justice efforts requires both devoting more resources to clean-ups in disadvantaged communities, and outreach to establish trust relationships. Not an easy task for an engineering and scientific organization. The increasing volume of plastics, with only a tiny amount recycled, and research that shows that the tiny particles from tire wear are very toxic to fish represent huge challenges that would best be met by establishing extended product liability legislation. We are learning that other modern chemicals, most notably those in the PFAS/PFOS family, represent new challenges.

Fortunately, in the Bay Area, we have had for nearly three decades a Regional Monitoring Program that uses high quality science to identify the highest priority water quality problems, and the progress we have made. This robust scientific effort is the envy of water quality staff in most of the country, and helps guide us to management approaches that will ensure water quality continues to improve.

Responding to climate change and rising seas represents another challenge that will require collaboration on an even larger scale. Along the shore of the East Bay we have several decades to implement measures that will limit the damage of increasing levels in the Bay. But no local government can go it alone, and coalitions between local governments, and with state agencies like Caltrans and State Parks, and regional agencies like the East Bay Regional Park District, will have to be built. My children and grandchildren will indeed live in interesting and challenging times. §

1941 — Foul and Offensive Discharges of Untreated Sewage

Prior to the passage of the Clean Water Act in 1972, untreated or minimally treated sewage flowed into the Bay from a multitude of discharge points. Treatment plant construction and consolidation began in the 1950s, and then really advanced in the mid- to late-1970s. The Berkeley Strawberry Creek channel carried untreated sewage that flowed directly into the Bay at the foot of University Avenue for over 50 years. In 1952, the East Bay Municipal Utility District (EBMUD) built a sewage treatment plant and the most of the sanitary sewer drains were re-routed to this facility.

> *"A sludge bank of considerable proportions has formed adjacent of the outlet, and foul odors, noticeable at all times, are particularly offensive at low tide on warm days. The water is black around the lumber wharves north of the outfall. Sewage solids are strewn for several hundred feet along the beach."*
>
> — Hyde et al. 1941

Photograph • Shira Bezalel, August 2022

2022
Valuable Habitat and Parkland

Like many of the sewage discharge points from the pre-CWA era, this area no longer receives municipal wastewater and has undergone redevelopment. The re-routing in 1952 did not entirely solve the problem — not all of the sewage inputs to the Creek were eliminated. In 1987 UC Berkeley created the Strawberry Creek Management Plan to address the still highly degraded conditions in the Creek. Implementation of the Plan from 1987 through 2004 led to greatly improved overall water quality in the Creek and downstream. Today there are no known connections between the sanitary sewer system and Strawberry Creek. Strawberry Creek now enters the Bay from a rectangular concrete culvert, south of University Avenue and west of the I-80/580 freeway. This area is now part of McLaughlin Eastshore State Park, managed by East Bay Regional Park District. The tidal flats at the creek mouth are habitat for shorebirds and other aquatic life, and popular for bird watching, dog walking, hiking, and biking. The San Francisco Bay Trail passes over the culvert.

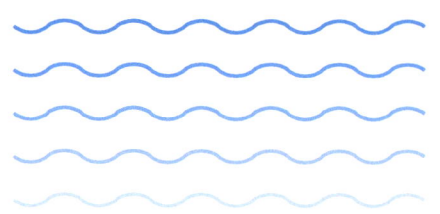

San Francisco Bay Regional Water Quality Control Board

Tom Mumley and Richard Looker

The federal Clean Water Act (CWA) established the structure and specificity for regulating pollutant discharges into waters of the U.S. and has served as the indispensable bulwark of water quality protection since its passage fifty years ago. The California Water Code, established 30 years prior to the CWA, supplements the CWA by providing California water quality regulators with strong and comprehensive regulatory authority to protect water quality. The 1972 CWA was expansive both in scope and breadth. Over the last 50 years, the San Francisco Bay Regional Water Quality Control Board (Water Board) has used CWA authority iteratively and adaptively in conjunction with its Water Code authority to resolve evolving water quality challenges and to progressively control sources of pollutants to the Bay and take other actions to protect, enhance, and restore water quality in the Region's waters. These efforts included use of authority provided in sections of the original CWA and the additional sections added in the 1987 CWA amendments. We provide an overview of the most relevant CWA sections and associated actions implemented by the Water Board in three time-periods below and provide our perspective on the challenges that lie ahead.

The CWA In Its Infancy (1972-1986)

The primary emphasis of both the Water Board and the CWA in the period following its passage was the establishment of water quality standards and implementation plans (Section 303), and control of wastewater discharges (Section 402).

The Water Board adopted the first water quality control plan for the San Francisco Bay Basin (Basin Plan) in 1975, which established water quality standards and implementation actions. Water quality standards consist of three core components. These include designated uses of a water body (e.g., recreation, aquatic life and wildlife habitat), water quality criteria (objectives in California terminology) to protect designated uses, and antidegradation requirements to protect existing uses and high quality/high value waters. It is noteworthy that California developed an antidegradation policy prior to the CWA via State Water Board Resolution No. 68-16, Statement of Policy with Respect to Maintaining High Quality of Waters In California, which remains an applicable and powerful policy today. The initial Basin Plan established water quality standards based on current science and information and focused implementation on waters adversely affected by wastewater discharges. However, it did recognize other sources and discharges of interests, and the 1982 Basin Plan updates included an initiative to prevent erosion and control sediment discharges from construction sites, in addition to enhanced wastewater initiatives.

The corresponding regulatory actions during this early period included regulation of municipal and industrial wastewater discharges and treatment facilities. Section 402 (which was added in 1977 CWA amendments) established the National Pollutant Discharge Elimination System (NPDES) program, and initial NPDES permits for municipal and industrial wastewater discharges (e.g., petroleum refineries) contained technology-based effluent limitations. For example, municipal facilities were required to meet discharge limitations based on secondary treatment (a combination of physical and biological treatment to remove biodegradable organics and suspended solids). These permits resulted in substantial reductions in pollutant loads to the Bay assisted significantly by about over $1 billion in grant funding provided by the CWA (Section 201) for municipal wastewater treatment systems in the region.

Another key addition to the CWA in the 1977 amendments was a list of toxic and priority pollutants (Section 307). Over time, EPA developed water quality criteria for pollutants on the list, which subsequently led to adoption of water quality objectives (WQOs) by the Water Board based on those criteria. In particular, as part of a substantial update to the Basin Plan in 1986, the Water Board established WQOs for some toxic pollutants, e.g., copper, mercury, nickel, and began to apply water quality-based effluent limitations (WQBELs) to wastewater discharge permits. Notably, the establishment of those WQOs and the ensuing WQBELs stimulated the creation of the RMP to assess attainment of the WQOs and to establish a scientific basis for the WQBELs.

Another key component of the 1986 Basin Plan update, based on recognition that attainment of WQOs would require control of pollutants in urban runoff in addition to wastewater, was the addition of a call for urban stormwater loading pollutant studies and management programs, starting with municipalities in Santa Clara County and, subsequently, Alameda County.

Powerful 1987 CWA Amendments (1987-1998)

The next chapter of CWA implementation begins with the 1987 Water Quality Act. This Act enhanced existing CWA sections and added new ones, expanding the scope of regulatory attention

Sewage outfall to the Bay in 1965 from the "No Deposit, No Return" documentary (*courtesy of KRON-TV and Bay Area TV Archive at SF State University*).

Courtesy of KRON-TV and Bay Area TV Archive, at SF State University

1965 Dumping Ground

Prior to the Clean Water Act and other environmental regulations of the 1970s, the Bay and its shoreline were a dumping ground for minimally treated sewage, industrial wastewater, polluted runoff, and solid waste.

Photograph · Shira Bezalel, August 2022

2022 Treasured Resource

The CWA contributed to a sea change toward protection and restoration of the Bay and the environment in general. However, the goals of the CWA have not yet been fully met.

to water quality issues and pollutant sources well beyond wastewater. The additions included Section 319, Nonpoint Source Management Programs, and Section 320, National Estuary Program. Section 402 was augmented to add municipal and industrial stormwater to the NPDES program and Section 304 was enhanced to add include additional requirements for certain NPDES permitted discharges associated with impaired waters.

Section 319 required each state to develop a Nonpoint Source Assessment Report to identify water bodies not meeting water quality standards due to nonpoint sources, and a Nonpoint Source Management Plan. Although the Water Board had, prior to the addition of Section 319 in 1987, given some attention to agricultural nonpoint sources, this was a timely call to action because it had become evident that control of municipal and industrial wastewater would not be sufficient to attain water quality standards in many water bodies. These nonpoint source efforts also stimulated enhanced attention to the list of impaired waters required by Section 303(d).

The National Estuary Program, created by Section 320 to protect and restore the water quality and ecological integrity of estuaries of national significance, included the San Francisco Estuary Project (now Partnership, SFEP), a collaboration of federal, state, and local agencies and non-governmental organizations working together to protect and restore water quality and the natural resources of the San Francisco Bay/Delta Estuary. Key to the SFEP's success has been the engagement of managers and scientists to build a common understanding of the state of science and identify management actions that would make a difference. SFEP provided a forum to build trust that prevails today among disparate and, previously, often adversarial entities. The consensus actions established through the initial Comprehensive Conservation and Management Plan in 1992 reflected a sea change in water quality and habitat protection and restoration efforts with substantial benefits. The development of the RMP was related to that effort, as well as the evolution of the Aquatic Habitat Institute to become the San Francisco Estuary Institute, which now provides a sustained forum to generate and assemble scientific information to inform management actions.

The 1987 amendments also added Section 304(I), Individual Control Strategies for Toxic Pollutants, which required states to identify waters that would not meet water quality standards for toxic pollutants after implementation of current effluent limitations. This section further required identification of specific NPDES permitted sources that were causing or contributing to the water quality impairment and to include control strategies in the NPDES permits to address the impairment. Consequently, the Water Board identified North San Francisco Bay water bodies as impaired by selenium due in part to petroleum refinery discharges and added requirements to the refinery NPDES permits to further control selenium discharges. The refineries implemented actions that reduced by more than half their combined selenium mass loading from the refineries, and subsequent monitoring by the RMP demonstrated that these load reductions eliminated the previously observed higher levels of selenium in the Bay in the vicinity of the refineries compared to other areas in the Bay.

The 1972 CWA established structure and specificity for regulating wastewater discharges. Similarly, the 1987 amendments, which added municipal and industrial stormwater discharges to the NPDES program via Section 402(p), provided structure and specificity to the Water Board's Urban Runoff Management Program established in the 1986 Basin Plan Amendments. The NPDES permit issued to municipalities in Santa Clara County for discharges of stormwater in 1990 was the first municipal stormwater NPDES permit issued in the US with pollutant control requirements. Similar permits for municipal stormwater discharges in Alameda County were issued in 1991 and for Contra Costa and San Mateo County municipalities in 1993. The need for information for future reissuances of these permits helped stimulate the formation of the RMP Sources, Pathways, and Loading Workgroup in 1998.

CWA and Water Board authority to regulate dredge and fill of water bodies has existed since the early CWA. Under Section 401, a federal agency may not issue a permit to conduct any activity that may result in any discharge into waters of the US unless a state certifies the discharge will comply with its water quality standards. Section 404 regulates the discharge of dredged or fill material into waters of the US, and the US Army Corps of Engineers issues permits for such discharges, which requires a 401-certification by the state. In the 1990s, the Water Board significantly increased its attention to these activities on its own initiative and in conjunction with the SFEP partners and stakeholders. This included enhanced protection of wetlands and streams affected by development and flood management projects and the emergence of the San Francisco Bay Long Term Management Strategy for Dredging and placement of dredged material. Yet again, the RMP was a valuable resource and provided Bay sediment quality data to inform management decisions.

The Modern Era (1999–Present)

The Water Board has adaptively improved its CWA regulatory programs during the last twenty plus years. Reissuance of wastewater NPDES permits, which had previously been a contentious process, has notably transitioned into uncontested Water Board actions over the last decade. This reduced conflict is a direct result of the ongoing trust between the Water Board and regulated entities and other stakeholders built through joint fact finding to establish the scientific foundation of these permits based on water quality data and information provided through the RMP. For municipal stormwater, in 2009, the Water Board issued for the first time a region-wide NPDES permit rather than issuing or reissuing individual city or county-based permits. This permit, reissued in 2015 and in 2022, implements a wide variety of water quality-based requirements for trash, copper, pathogens, pesticides, mercury, and PCBs. The RMP is essential to gathering data and answering management questions associated with many of these contaminants.

Section 303(d) has become one of the most impactful parts of the CWA guiding the Water Board's regulatory efforts. It requires states to identify water bodies that do not attain water quality standards and to develop allowable total maximum daily loads (TMDLs) for pollutants causing those impairments. Although the Section 303(d) TMDL requirements existed in the original CWA, successful citizen's lawsuits against EPA in the late 1990s for lack of TMDLs in other parts of the US motivated California Water Boards to ramp up TMDL development. The RMP's Status and Trends component provides most of the water quality data to determine if the Bay is meeting standards, and both RMP-funded special studies as well as collaborative discharger-funded studies helped inform the TMDLs and implementation plans. The Water Board established TMDLs for mercury and PCBs in San Francisco Bay in 2008 and 2010 (effective dates) and for selenium in North San Francisco Bay segments in 2016. The Water Board continues to work

> *This reduced conflict is a direct result of the ongoing trust between the Water Board and regulated entities and other stakeholders built through joint fact finding to establish the scientific foundation of these permits based on water quality data and information provided through the RMP*

with RMP scientists and stakeholders to address key management information needs associated with implementing these TMDLs.

The Water Board adopted site-specific water quality objectives for copper and nickel to resolve 303(d) listings for San Francisco Bay, south of the Dumbarton Bridge in 2003 and for the rest of the Bay in 2007. Data generated by the RMP played a key role in these efforts, and most importantly, the RMP provided a reliable means to track and demonstrate that with the loading of copper and nickel allowed by WQBELs based on the new objectives, SF Bay quality would not degrade (and it hasn't). The Water Board also adopted site-specific water quality objectives for cyanide in 2007, not because the Bay was impaired, but because WQBELs based on the existing water quality objective posed expensive compliance challenges for wastewater discharges while RMP data indicated beneficial uses of the Bay were not impacted. As for copper, the RMP provided a sustainable means to track levels of cyanide in the Bay as well as potential degradation of Bay waters.

The San Francisco Estuary Partnership is now a mature collaborative that updated its original 1992 management plan in 2007, 2016 (renaming it the Estuary Blueprint), and 2022. The 2022 Estuary Blueprint established an updated list of 25 priority actions that should be implemented by partner agencies, including the Water Board, for restoring the health of the Estuary's chemical, physical, and biological processes. There are priority actions associated with promoting resilience and adaptation to climate change and sea level rise, aquatic resource protection, tidal marsh restoration, restoring watershed connections to the Estuary, sediment management, and addressing emerging contaminants, trash, and nutrients. The RMP continues to inform and help accomplish these actions by providing both monitoring data as well as technical information through special studies.

Long-term monitoring data generated by the RMP in partnership with the USGS suggest that the Bay may be becoming less resilient to nutrients, which could lead to lower dissolved oxygen concentrations as well as harmful algal blooms. Wastewater discharges constitute the largest single source of nutrients so appropriate regulation of nutrient concentrations in wastewater effluent will be part of the solution. However, in 2012, rather than imposing more stringent wastewater discharge requirements, the Water Board in collaboration with the municipal wastewater agencies initiated a Nutrient Management Strategy that provided substantial funding beyond their RMP contributions to support a nutrient science program to inform nutrient management actions. In addition, the municipal wastewater agencies have committed to evaluate wastewater treatment options for reducing nutrient loading to the Bay and proactively implement identified technologies and strategies as possible. These treatment options include enhanced conventional treatment technologies, nature-based systems, (e.g., treatment wetlands), and recycling of treated wastewater. The Water Board has recognized these collaborative actions via a Bay-wide watershed permit covering all municipal wastewater nutrient discharges, first issued in 2014 and reissued in 2019.

The Water Board has continued to increase its attention to dredge and fill activities, and its regulatory efforts have risen in prominence. While protecting streams and wetlands remains a priority, the Water Board has progressively put more effort towards permitting of restoration projects and opportunities for beneficial reuse of dredged sediment driven by both the needs of today current and future challenges posed by climate change and sea level rise. This has led to further scientific information needs to inform and track restoration efforts and to understand sediment fate and transport in the Bay and its interface with its margins. Here again the RMP is playing a key role via its increased focus on sediment sources and transport, and the long-awaited creation of a Wetlands Regional Monitoring Program is at hand.

The Next Chapter — Adapting to Complex Challenges

As we look to the future, we are already confronting two difficult problems that are likely here to stay, contaminants of emerging concern (CECs) and climate change. These two challenges generate a great number of scientific information needs for which the RMP will, yet again, play a critical role. CECs and climate change also present myriad regulatory challenges that will require innovative application of the CWA along with state laws and regulations.

CECs pose multiple challenges. There are more than 100,000 chemicals registered or approved for commercial use in the US, and we typically have limited information concerning the environmental risks posed by them. Through the RMP, we have made progress gathering occurrence, fate, and toxicity data for many of these chemicals, e.g., pesticides, flame retardants, per- and poly-fluorinated alkyl substances (aka PFAS), bisphenols, and

> *As we look to the future, we are already confronting two difficult problems that are likely here to stay, contaminants of emerging concern (CECs) and climate change*

surfactants, which gives us cause to consider a new priority pollutants list to supplement or replace the archaic list from the late 1970s. In line with that consideration, much of the attention of the RMP has shifted from legacy contaminants to CECs. We also want to avoid the "regrettable replacement" dilemma, wherein a product ban or phase-out of a particular CEC results in a replacement chemical that is also toxic. For example, bisphenol-A in many plastic products has been replaced with bisphenol-S, which is also toxic and now found in the Bay.

Climate change presents an enormous variety of societal and regulatory challenges, but here we also find some opportunities for solutions that not only help solve water quality problems but can simultaneously enhance the Bay's climate change resilience. Achieving these outcomes will require creative and nimble use of the CWA, particularly in regulating wastewater discharges, managing sediment, and in enhancing and restoring streams and wetlands, which provide vital ecosystem services that are even more important in promoting climate change resilience.

The nutrient management challenge offers us a golden opportunity to employ strategies that help control nutrients, and some CECs, while simultaneously improving the Bay's resilience to climate change and providing critical shoreline marsh habitat. Building more advanced wastewater treatment plants to substantially improve nutrient removal would cost more than $10 billion. If we must invest that much money, we must consider all possibilities for beneficial reuse of treated wastewater. One multiple benefit option is the horizontal levee. The sloped (wetland) vegetation of this nature-based solution reduces the adverse impacts to shorelines from coastal flooding, storm surge, and wave action. The natural bacterial and biological processes in the wetland system remove nutrients and break down contaminants, including some CECs. The wetland soils filter contaminants from the water flowing through, thus providing additional treatment for wastewater effluent before discharge to the Bay. These systems also provide habitat and recreational opportunities associated with marshes and other coastal habitats. There's even a potential extra benefit in that recent studies indicate horizontal levees may be an effective means of reducing nutrients and contaminants found in the concentrated wastes generated by the reverse osmosis process used in recycled water advanced purification treatment systems.

A second climate change-related water quality regulatory challenge is how to manage sediment intelligently. The resilience of San Francisco Bay shore habitats, such as tidal marshes and mudflats, is essential to all who live in the Bay Area. These baylands protect billions of dollars of bay-front housing and infrastructure (including neighborhoods, business parks, highways, sewage treatment plants, and landfills). However, climate change poses a great threat, because there may not be enough natural sediment supply for tidal marshes and mudflats to gain elevation fast enough to keep pace with sea-level rise.

Historically, sediment would have been washed down watersheds to the Bay, accreting in wetlands and replenishing sediment lost to tidal action. After centuries of development, however, the Bay's wetlands have largely lost their connection to sediment sources, leaving them increasingly vulnerable to inundation as sea levels rise. We need to fine tune our regulatory strategies for flood control projects and creek restoration in order to re-establish the connectivity between watershed sediment supply and wetlands. The RMP Sediment Workgroup is working on providing technical information for sound management and regulation. However, policy changes in the Basin Plan and other federal and state authorities may be required in order to optimize the use of dredged sediment for wetland creation and restoration in an intelligent and environmentally responsible manner. §

Courtesy of KRON-TV and Bay Area TV Archive at SF State University

1965 — Unsightly and Unhealthy

Aquatic Park in San Francisco has been frequented by swimmers since the 1800s, as evidenced by the long-term presence in this area of two sporting clubs that include swimming: the Dolphin Club (present since the 1800s) and the South End Rowing Club (present since the early 1900s). Although quantitative monitoring of trash abundance in the Bay has not been performed, it is safe to say that in the pre-1970s era of dumping of untreated sewage and solid waste in and around the Bay, bacterial contamination and trash were bigger problems than they are today.

Photograph · Shira Bezalel, September 2022

2022 Improved, But Not Pristine

Aquatic Park remains one of the Bay's most treasured swimming spots. While there is undoubtedly less trash in the Bay now than 50 years ago, trash is still included on the CWA list of problem Bay contaminants, both as an aesthetic nuisance and a threat to aquatic life. The Water Board has promulgated aggressive goals for removing trash from urban stormwater, and hundreds of millions of dollars have been spent on trash capture devices and other efforts to remove trash before it gets to the Bay. For bacterial contamination, Aquatic Park is relatively clean, consistently receiving "A" beach report card grades during the summer months. During wet weather in the winter, however, like many Bay locations, the grades are lower. In 2021, the two Aquatic Park locations monitored got a "B" and a "C". This is relatively good for wet weather grades though — all seven other San Francisco beaches monitored got an "F."

US Environmental Protection Agency

Tomás Torres

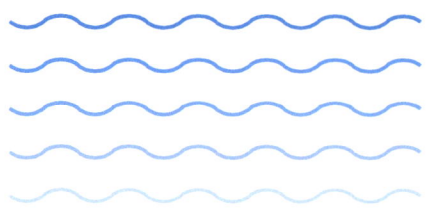

The Clean Water Act and EPA's San Francisco Bay Water Quality Improvement Fund

The Clean Water Act (CWA), when written in the 1970s, had a section for every conceivable way to start controlling pollutants and restoring waterbodies. Those sections brought (and bought!) us wastewater treatment plants, drinking water treatment plants and controls on industrial discharges to recover and sustain thriving waterbodies and aquatic ecosystems. There was also a lesser known yet influential part of the CWA Section 320 that would enable steady and non-regulatory water improvements through cooperative partnerships. CWA Section 320 authorized the formation of the San Francisco Estuary Partnership after recognizing the San Francisco Bay-Delta Estuary as a nationally significant estuary. It then supported the release of a regional restoration plan called the Estuary Blueprint which propelled the need to improve the science and soon after the San Francisco Estuary Institute was created. In 2008, EPA began using CWA Section 320 to invest federal grant dollars in San Francisco Bay to accelerate wetlands and water quality restoration known as the San Francisco Bay Water Quality Improvement Fund (SFBWQIF). To date, more than $71.4 million in EPA grants have been awarded to 80 different partners.

Grant recipients have represented a broad network of partners across the nine Bay Area counties and their work at the local and regional level has resulted in numerous CWA success stories. The diversity of projects the SFBWQIF can fund is one of its greatest strengths; examples include helping cities to improve stormwater quality by upgrading urban infrastructure from "gray to green"; supporting cities and counties in their efforts to find innovative solutions to reduce human-related water quality impacts, including communities experiencing homelessness; and reducing levels of trash and microplastics in stormwater and Bay waters. Other projects have accelerated the restoration of watersheds impacted by wildfires; facilitated the beneficial reuse of dredged material for construction projects, including wetlands restoration; constructed multi-benefit shoreline projects that restore habitat while providing flood protection and pollutant reduction; and reduced nutrient inputs into San Francisco Bay through piloting of new removal technology at wastewater treatment plants.

Today, the most challenging water pollutants including sediment, mercury, nutrients, pesticides, pathogens, and polychlorinated biphenyls (PCBs) are continuing to be addressed in San Francisco Bay. Under the CWA, and through our state partner the San Francisco Bay Regional Water Quality Control Board, many of these pollutants now have a TMDL (total maximum daily load) or "pollution diet," to compel actions that reduce these pollutants over time. The ability to fund actions required by permits is another important flexibility and strength of the SFBWQIF as many other federal funding sources are not able to do so. Consequently, over $31 million from the SFBWQIF has been invested in projects to implement TMDL and stormwater permit actions.

Now, more than ever, the protection and restoration of tidal wetlands are seen as essential measures to preventing the loss of our shorelines and critical infrastructure as sea levels continue to rise. The SFBWQIF has also been an important tool to leverage additional funding streams to benefit wetlands restoration. EPA's investment of nearly $32 million through the SFBWQIF has benefited 6,700 acres of wetlands around the Bay, adding to state and regional funding that have brought over 20,000 acres into active tidal restoration. These efforts are a positive step towards reaching the "Baylands Goal," set in 1999, of restoring 100,000 acres of tidal wetlands in San Francisco Bay. EPA's grant funding through the CWA will continue to be instrumental in supporting tidal marsh restoration that, when mature, will help stem the impacts of sea level rise.

Today, as we commemorate the 50th year of the Clean Water Act, it is fitting that we also celebrate an unprecedented increase in funding by Congress for the SFBWQIF, now totaling $29 million as part of the Bipartisan Infrastructure Law and base appropriations. These funds will further support our ability to accelerate progress in the face of climate change stressors and the needs of underserved communities around San Francisco Bay for the next generation. §

San Francisco Bay Water Quality Improvement Fund

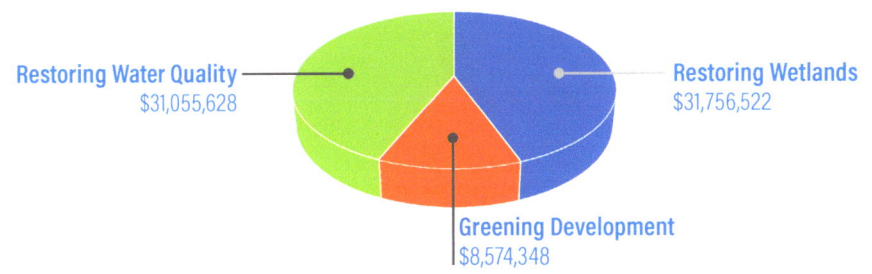

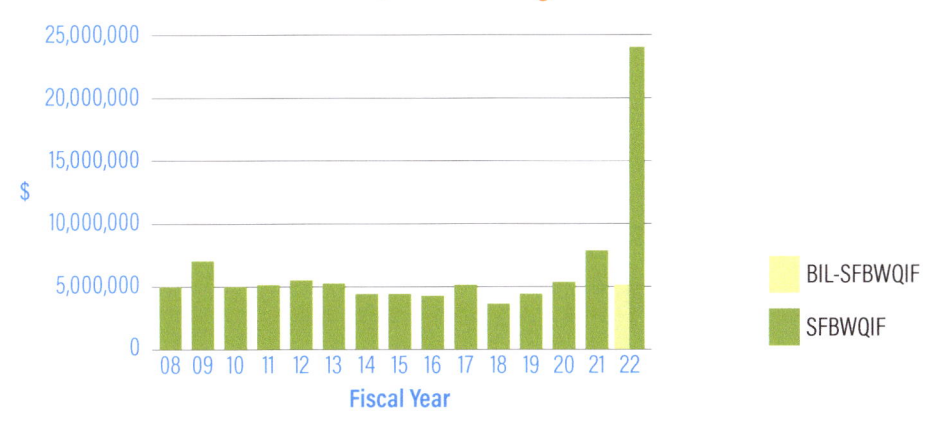

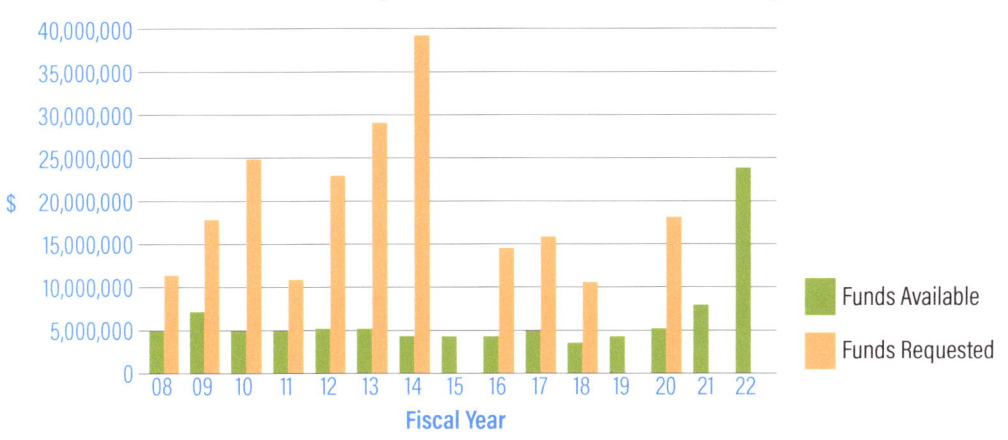

Photograph • Belinda Rains, 1972, from USEPA's Documerica Project

1972 Partially Treated Sewage Discharged from Many Locations

Prior to the CWA of 1972, sewage was discharged into shallow Bay waters at many points along the shoreline. A USEPA project that ran from 1972-1977 called Documerica captured images of polluted areas across the country, including this one of partially-treated sewage entering the Bay from the Union Sanitary District (USD) Biofiltration Plant in 1972. The USD had three regional plants at that time, and this outfall was from the Irvington Plant, the only one of the three with biofiltration, which is one component of secondary treatment.

Photograph • Shira Bezalel, August 2022

2022 Upgrades, Consolidation, and Shoreline Restoration

Between 1972 and 1987, the CWA provided $1.2 billion to support Bay Area wastewater treatment upgrades and consolidation of systems and outfalls. In 1981, the main treatment plant for the Union Sanitary District was completed (serving Fremont, Newark, and Union City), with the treated effluent discharged through an outfall shared with other East Bay municipalities to deep water, away from the Bay shoreline. In 1982 the culvert and outfall structure for the Irvington plant were demolished and the levees were restored to their pre-outfall condition.

Today this area is recognized as valuable wetland habitat that has either been included or is proposed for inclusion in the South Bay Salt Pond Restoration Project.

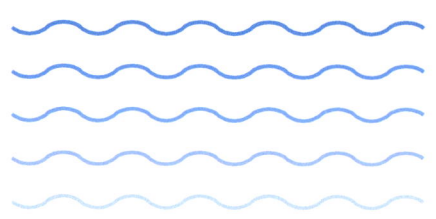

BACWA

Lorien Fono

Bay Area Clean Water Agencies (BACWA) is a joint powers agency whose members own and operate publicly-owned treatment works (POTWs) and sanitary sewer systems that collectively provide sanitary services to over 7.1 million people in the nine-county San Francisco Bay Area. BACWA members are public agencies, governed by elected officials and managed by professionals who protect the environment and public health.

The Clean Water Act serves as the legal framework for how the clean water community views our mission. The National Pollutant Discharge Elimination System (NPDES) program established by the Clean Water Act provides the regulatory structure for wastewater treatment plants to comply with discharge standards and prohibitions. Since the adoption of the Clean Water Act, investments in clean water infrastructure have resulted in dramatically improved water quality in San Francisco Bay, even in the face of a near-doubling of regional population. Most of our facilities were built with federal grant funding that came with the adoption of the Clean Water Act.

While the Clean Water Act is foundational, the Bay Area clean water community has extended its vision beyond the minimum requirements of the law to broaden what it means to be stewards of the environment and the community in three major ways.

We Are Responsive to a Changing Environment

When the Clean Water Act was adopted, its framers did not envision the need to adapt to a changing climate. On the surface, this adaptation means protecting our low-lying facilities from sea level rise; adjusting our treatment processes to adapt to water conservation, which reduces sanitary sewer flows and produces more concentrated wastewater; and adjusting to changing precipitation patterns. Our ability to continue to operate under these new conditions is not our sole concern. We see our responsibility to our communities as extending beyond our fence lines, and our charge to protect our ecosystem as extending beyond the area of influence of our outfalls.

This recognition of climate change impacts is driving our community to form partnerships that will allow us to play a role in both climate change adaptation and mitigation. Drought is driving an increase in recycled water production to improve our region's water supply resiliency. According to a BACWA recycled water study now in progress, wastewater agencies in our region are expected to more than double their recycled water production over the next 20 years. Wastewater agencies in the Bay Area are also planning and implementing treatment wetlands that can improve effluent quality prior to discharge, while simultaneously protecting upland infrastructure and enhancing habitat.

Climate challenges and opportunities extend beyond just our traditional clean water mandate. The State is looking to POTWs as critical infrastructure to receive green waste and food waste from our communities, transform them into energy via co-digestion, and find beneficial end uses for the resulting biosolids that also sequester carbon. The clean water community is thus assuming the cross-disciplinary tasks of water resources, carbon, and energy management. We are taking on these new responsibilities while also controlling air emissions that primarily impact the often overburdened communities that live near our facilities.

We Engage in Proactive Joint Fact Finding with Regulators and the Science Community to Inform Water Quality Management

When the Clean Water Act was first implemented in the 1970s, the focus was controlling discharges of suspended solids, organic material that consumes oxygen, and industrial pollutants. Over time, NPDES permits have come to regulate many additional toxic pollutants, such as solvents and legacy pesticides. Our agencies have been extremely successful at meeting the effluent standards introduced under this framework. However, the major ecosystem challenges of our time keep evolving, requiring a knowledge of local impacts. While nutrients have garnered much attention on a national level, gaining a nuanced understanding about the impact of nutrients on San Francisco Bay ecosystems requires a regional approach. Regionally relevant decision-making has been the focus of the Nutrient Management Strategy (NMS), a collaborative made up of regulators, wastewater agencies, and other regional stakeholders. The NMS provides a structure for allocating funding, largely provided via the POTW community, to best inform Bay nutrient policy. The NMS serves as a national model of multi-stakeholder collaboration for guiding science to make management decisions.

The regional clean water community has also been proactive at considering the emerging contaminants that are not yet regulated through the Clean Water Act. We work closely with the Regional Monitoring Program (RMP) to examine wastewater as a possible pathway to the Bay for pesticides, pharmaceuticals, microplastics, and other trace constituents. We have worked with the Regional Water Board to provide a sustainable funding source for the RMP's emerging contaminants program. In 2020, the State Water Board issued a blanket investigative order to POTWs throughout the state to conduct monitoring of per- and polyfluorinated alkyl substances (PFAS) in influent, effluent, and biosolids. Because of the success of our collaboration with the RMP, the State Water Board allowed the POTWs in our region to perform a special study to monitor at representative facilities and investigate the sources of PFAS to our facilities, in lieu of the blanket monitoring requirements that were issued to POTWs in other regions. This example illustrates how our track record in collaboration and support for science-based decision-making allow us to take a more targeted, hypothesis-driven approach to today's major environmental questions.

We Responsibly Serve Within Our Communities

The Clean Water Act charges POTWs to serve as stewards of water resources, but our communities expect that we take a more holistic view of our role in society. One illustration of this mission is the way agencies in our region stepped up to provide samples for wastewater surveillance to inform public health management during the COVID-19 pandemic. This service falls outside of our traditional clean water role, but has been a key feature of pandemic response.

With enhanced focus on environmental justice, the clean water community has been taking a hard look at our role within the communities we serve. Historically, our direct engagement with our communities has been focused on protecting public health by keeping the public away from sewers and wastewater. We also engage in public education to support pollution prevention and protect subsistence fishing populations from contaminants such as mercury and PCBs that bioaccumulate in fish. While this continues to be important work, it has become increasingly clear that our strategies for decision-making impacting the communities we serve must be re-envisioned. A major charge of this decade will to develop relationships with historically marginalized communities to incorporate diverse voices into planning.

The Clean Water Act will continue to serve as a foundation for our operations and long-term planning. The Bay region's clean water community views its mission as extending beyond the minimum requirements of the Clean Water Act in how we serve our neighborhoods and ecosystems. The past 50 years have brought significant evolution in our role, and we expect the next 50 years to continue and accelerate that trend. §

The Evolution of the Bay Area's Largest Wastewater Treatment Plant

1956

1956
The San José-Santa Clara Regional Wastewater Facility began as a primary treatment plant surrounded by farmland. It was constructed and initiated operation in 1956-1957, serving a population of 380,000. Primary treatment uses physical processes to remove settleable and floating fats, oils, and grease. Since primary treatment does not effectively remove organic matter, dissolved oxygen in the receiving water (Artesian Slough) was near zero much of the time.

1964
The first major upgrade was implementation of secondary treatment in 1964. In secondary treatment air is pumped into the wastewater to nurture the growth of naturally occurring aerobic bacteria that metabolize organic matter. This greatly reduced the amount of organic matter discharged to Artesian Slough and dissolved oxygen levels in the Slough improved considerably.

1979

The second major upgrade was addition of nitrification and filtration facilities in February 1979. These additional process steps further reduced the oxygen demand of the treated effluent, and further improved dissolved oxygen in Lower South Bay.

1997

The third significant upgrade was modification to Biological Nutrient Removal (BNR) in late 1997 and early 1998. BNR removes nitrogen and phosphorus through the use of microorganisms under aerobic and anaerobic conditions in the treatment process. This sharply reduced the overall nitrogen loading to the Lower South Bay, driving a 40% decrease in nitrogen concentrations in the Lower South Bay and demonstrating the potential effectiveness of load reductions.

2022

The San José-Santa Clara Regional Wastewater Facility is the largest municipal wastewater treatment plant (WWTP) in the Bay Area and the largest advanced WWTP in the western US. It serves a population of 1.5 million. The facility is in the midst of planning another major upgrade. A Master Plan was completed in 2013 after several years of study and evaluation of the technology, infrastructure condition, and projections of future needs for the service that the Facility provides to businesses and residents. An approved Capital Improvement Plan (CIP) envisions over $2 billion of investment, upgrades, repairs, replacements, and improvements over a 30-year span. The CIP is in its first phase, which is envisioned to be a 10-year effort with $1.4 billion of improvements that will touch virtually every aspect of treatment operations.

Bay Area Municipal Stormwater Agencies

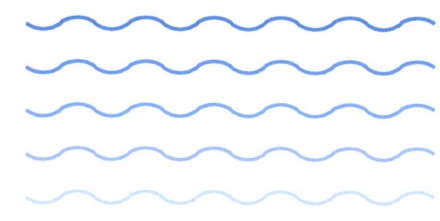

Chris Sommers and Jill Bicknell · Santa Clara Valley Urban Runoff Pollution Prevention Program
Reid Bogert · San Mateo Countywide Water Pollution Prevention Program
Emily Corwin · Solano Stormwater Alliance

The Bay Area Municipal Stormwater (BAMS) Collaborative is an informal association comprised of the municipal stormwater programs in the San Francisco Bay Area, which represent over 100 public agencies, including cities and towns, counties, and special districts. The BAMS Collaborative focuses on addressing regional challenges and opportunities to improve the quality of stormwater flowing to our local creeks, the Delta, San Francisco Bay, and the Pacific Ocean. The BAMS Collaborative represents and advocates for common interests of member programs at the regional and state levels.

The 1972 Water Pollution Control Act (commonly referred to as the federal Clean Water Act or CWA) and its amendments comprise some of the most expansive and foundational environmental laws enacted to date at the federal level. The CWA sets the overall goals of protecting and restoring the physical, chemical, and biological integrity of freshwater ecosystems (e.g., small creeks, large rivers, and lakes) and coastal waterways and wetlands, including San Francisco Bay and the Pacific Ocean. The integrity of these surface waters can be impacted by a number of factors, including pollutants in rainfall that flows over impervious surfaces, such as paved streets, parking lots, and building rooftops, and is conveyed through stormwater drainage systems, largely untreated, to surface waters. In the Bay Area, stormwater drainage systems are largely comprised of a complex array of storm drain inlets, underground pipes, and open channels, generally owned and operated by public stormwater agencies (i.e., cities, counties and special districts).

The National Pollutant Discharge Elimination System (NPDES) permitting program provides the US Environmental Protection Agency (USEPA) (or states designated by USEPA) the authority to regulate applicable "discharges" to surface waters, including those conveyed through stormwater drainage systems. In the Bay Area, CWA requirements have been implemented though NPDES permits issued by the San Francisco Bay Regional Water Quality Control Board (Water Board) and requirements included in these permits have continued to evolve to address sources of legacy pollutants such as PCBs and mercury that are impacting the Bay and contaminants of emerging concern that may be on the verge of or already causing impacts to surface waters in ways that are not yet well understood. NPDES permit requirements issued in the Bay region have set an increasingly high bar for stormwater quality management by local public agencies with no federal or state funding to support these programs. Even so, local public stormwater agencies have made considerable progress in protecting stormwater quality and reducing pollutant impacts. These agencies, however, face the challenge of controlling pollutants in stormwater without dedicated or adequate funding sources while stormwater infrastructure is nearing its life expectancy and climate change continues to contribute to drainage problems and flooding during larger storms. With the 50th anniversary of the CWA, it is important to reflect on the history of stormwater management in the Bay Area, including the successes achieved, but more importantly to provide context to the ongoing and new challenges that public stormwater agencies will face over the next 50 years.

As groundbreaking as the original CWA was, it largely focused on addressing discharges of wastewater from publicly-owned sewage treatment plants and industrial facilities. Not until 15 years after its adoption, with the passing of the Water Quality Act of 1987, was the federal CWA expanded to address discharges from stormwater conveyance systems (i.e., storm sewer systems that are separate from sanitary sewers). With the 1987 Amendments, Section 402 of the CWA was expanded to establish the foundation for regulating discharges from municipally-owned stormwater conveyance systems, which are commonly referred to as "MS4s." Although regulated as a "point source" type of discharge like wastewater treatment plants, Section 402(p) of the CWA established a distinctly different type of regulatory framework for discharges from MS4s—one based on reducing the discharge of pollutants to the Maximum Extent Practicable (MEP), rather than to a numeric endpoint. The MEP approach was selected largely due to the inherent complexities and challenges of controlling the quality of rainwater that flows intermittently over the surfaces of urban watersheds, through a distributed underground system of interconnected pipes, and into local waterways from thousands of "outfalls." By including the MEP concept, the US Congress showed that it understood that controlling pollutants conveyed through MS4s is not as straightforward as sending wastewater from homes and businesses to a single point of treatment.

In addition to expanding the scope of water quality regulations, the 1987 CWA Amendments also established federal funding to conduct the National Urban Runoff Program (NURP), an expansive science and engineering program that ran until the early 1990s. Studies conducted by NURP largely focused on characterizing stormwater runoff through water quality monitoring and conducting stormwater control measure investigations to evaluate their effectiveness. Stormwater controls generally include three types of actions: true source controls (e.g., pollution prevention policies, ordinances, or laws), institutional source controls (e.g., street sweeping and debris removal), and treatment controls (e.g., mechanical systems, low impact development, and green stormwater infrastructure). The Bay Area served as a petri dish for many of the NURP studies, whose results helped inform the requirements included in the first NPDES stormwater permits issued by Water Board to local public agencies in the early 1990s. Cities, counties, and flood control districts in Santa Clara and Alameda counties received the first NPDES stormwater permits in the nation and soon after, public agencies in San Mateo and Contra Costa counties were also issued permits.

> *Cities, counties, and flood control districts in Santa Clara and Alameda counties received the first NPDES stormwater permits in the nation*

Because NPDES stormwater permits were issued on a county-by-county basis, Bay Area public agencies organized themselves into countywide stormwater management programs, which provided opportunities for public agencies to collaborate and establish consistency in management approaches. These countywide programs eventually formed the Bay Area Stormwater Management Agencies Association (BASMAA), which recently transformed into the Bay Area

▲ Before and after installation of green infrastructure in El Cerrito
(Photographs by Sarah Pearce).

Municipal Stormwater (BAMS) Collaborative - an informal collaboration of the same entities. These organizations fostered further collaboration among Bay Area public agencies and provided a single access point for Water Board staff to engage with Bay Area stormwater managers. The countywide programs in coordination with BASMAA took a lead role in the development, coordination, and streamlining of novel and award-winning stormwater management approaches in the Bay Area over the last three decades, including the development of the seminal Start at the Source design manual in 1999, which spearheaded the Low Impact Design (LID) and Green Stormwater Infrastructure (GSI) movement in the Bay Area. These and other approaches championed by public stormwater agencies have proven to be foundational for addressing new challenges in the Bay Area and have provided a model for other regions in California and throughout the US.

The first NPDES stormwater permits issued by the Water Board included a relatively small set of control programs that were generally focused on improving stormwater quality, not necessarily on addressing a single pollutant or set of pollutants. These initial controls included enhancing existing street sweeping to improve the interception of sediment-associated pollutants from these impervious surfaces; preparing design guidelines for new and redeveloped properties to minimize impervious surfaces and promote on-site infiltration; developing and implementing illegal dumping and illicit discharge response programs so that cities and counties could effectively respond to pollutant spills; and creating public education and outreach campaigns to raise public awareness about watershed protection and improve behaviors that impact stormwater quality. Over the past three decades, these "core" stormwater controls have continued to evolve and mature into the model programs they are today.

During the same timeframe that the first NPDES stormwater permits were being issued in the Bay Area, public stormwater agencies also began establishing and participating in new monitoring programs to better understand the quality of stormwater entering local creeks, rivers, and the Bay. With the support of local public agencies (stormwater and wastewater), the Regional Monitoring Program for Water Quality in San Francisco Bay (RMP) was also formed around this time to assess the status, trends, and sources of Bay pollutants. Data collected from these monitoring programs soon led to several "impairment" listings by the Water Board under Section 303(d) of the CWA, which in turn spawned the development of pollutant-specific water quality restoration programs by the Water Board called Total Maximum Daily Loads (TMDLs). Stormwater and Bay monitoring and modeling conducted collaboratively through the RMP, with guidance from the RMP's newly formed Sources, Pathways, and Loadings Workgroup (SPLWG), helped stakeholders better understand the importance of different pollutant pathways, including stormwater.

As a result of RMP and local agency data collection, stormwater drainage systems were identified by the early 2000s as important pathways of copper, nickel, polychlorinated biphenyls (PCBs), and mercury to the Bay. Additionally, monitoring conducted in local tributaries by cities, counties, and the Water Board also indicated that urban creeks and rivers in the Bay Area were impaired by pesticides and trash. The 303(d) listings and TMDLs provided the impetus for the Water Board to include pollutant-specific requirements in reissued NPDES stormwater permits beginning in the early 2000s. These new requirements created a shift in the focus of stormwater management in the Bay Area – adding to the core programs requirements that focused on controls that would address specific pollutants identified as impacting the beneficial uses of the Bay and local creeks and rivers.

Since these pollutant-specific requirements (as well as the core programs) were generally applicable to all stormwater discharges to Bay Area urban creeks, rivers, and the Bay, the impetus was created for the Water Board to issue the first regional NPDES stormwater permit in 2009, which has come to be known as the Municipal Regional Permit (MRP). The issuance of the MRP not only furthered the need for collaboration among local public agencies on stormwater control measure planning and implementation, but also established a greater need for coordination at a regional scale on monitoring being conducted by the countywide stormwater programs and the RMP. Coordinated monitoring was instrumental to informing stormwater management actions needed to address early water quality objectives for copper and nickel adopted for the Bay (south of the Dumbarton Bridge in 2003 and for the rest of the Bay in 2007) and pollutant reductions required by the PCBs (2008) and Mercury (2009) TMDLs. In response, countywide stormwater programs, the Water Board, and San Francisco Estuary Institute staff formed the RMP Small Tributaries Loading Strategy (STLS) Team to help develop and implement new monitoring strategies to address stormwater information gaps related to these pollutants. The monitoring strategy formed by the STLS team helped guide stormwater data collection and watershed modeling over the next decade, ultimately leading to more accurate estimates of pollutant loading and reductions achieved to date through existing controls, as well as of the extent and magnitude of additional control measure implementation needed to address the TMDL goals for stormwater. After collection of hundreds of stormwater and sediment samples and many analyses and modeling runs, our collective understanding of the spatial and temporal aspects of copper, nickel, PCBs, mercury, and other pollutants in stormwater and the transport processes to the Bay from stormwater conveyances is at an all-time high, providing a strong foundation for further refinements to monitoring approaches, watershed models, and analyses focused on new and emerging contaminants potentially impacting the Bay.

Over the last 35 years, the ongoing collaboration between the RMP and stormwater agencies has also resulted in identification and implementation of tangible and significant stormwater management actions that have improved (or will improve) the quality of surface waters in the Bay Area for years to come. One example is the monitoring dataset collected by the RMP in 2010, which supported the adoption of Senate Bill (SB) 346, also known as the California Motor Vehicle Brake Friction Material Law. SB 346 was developed through the Brake Pad Partnership, a collaboration of stormwater agencies, brake manufacturers, water quality regulators, and environmental groups. The group worked together for over a decade to determine if copper from brake pads was a significant contributor to stormwater and the Bay and evaluate the management options for controlling this source. The RMP data informed a complex set of discussions among these entities, ultimately resulting in a version of a bill that would protect stormwater quality, meet NPDES permit requirements, and give the industry the time and flexibility needed to develop, test, and produce alternative brake pad materials. Another example is the collective efforts of the RMP and stormwater agencies to identify and inform the abatement of PCB sources associated with older industrial areas adjacent to the Bay. Over the last decade, the RMP and countywide stormwater programs have effectively coordinated their monitoring efforts to identify specific drainages that are likely contributing significant levels of PCBs to the Bay via stormwater. Stormwater programs have used these data to conduct more refined source investigations that have led to the identification of specific PCB source areas in local watersheds. Many of these areas have been referred to the Water Board or USEPA for further investigation and abatement, leading to significant reductions of PCB loads to the Bay. These are just a couple of ways in which a concerted monitoring effort through the RMP has supported our collective progress towards CWA goals over the past several decades.

> *The future of stormwater management over the next 50 years is a bright star on the constellation of water quality improvement, but must continue to be elevated with adequate funding and political support*

Although there have been many achievements over the last 35 years to improve stormwater quality in the Bay Area, local public stormwater agencies face significant ongoing and future challenges, potentially impeding their ability to help achieve the protection and restoration goals of the CWA. These challenges include, but are not limited to: 1) a lack of dedicated and adequate funding for stormwater management and an aging infrastructure system; 2) the onslaught of new and emerging contaminants that are more ubiquitous in the urban environment and more challenging to control once in stormwater; and 3) the impacts of climate change and sea level rise. With regard to funding, municipal stormwater systems are public facilities, but they differ from other public utilities such as water, sewer, and garbage in one key aspect: other utilities existed prior to the passage of constraining state legislation and are financially supported by service fees. By comparison, many stormwater agencies rely on the public agency's general fund, which presents a major challenge for elected officials as they balance the funding of stormwater management with other programs supported by the general fund, including law enforcement, fire protection, paramedics, parks, street lighting, and libraries. The lack of adequate and dedicated funding sources for stormwater management is exacerbated by the relatively new issues of emerging contaminants and climate change. Over the last 35 years, Bay Area stormwater agencies have spent hundreds of millions, if not billions, on successfully intercepting and removing pollutants like copper, PCBs, mercury, pesticides, and trash from stormwater. New and emerging contaminants, however, such as PFAS and microplastics pose a significant concern to stormwater agencies, and the approaches used to address legacy and current types of pollutants are not likely the most cost-effective and sustainable approaches for these new types of contaminants. Stormwater agencies, with the assistance of the RMP, are collecting data to support more proactive measures to address these emerging contaminants before they impact surface water quality. Eliminating the generation of pollutants through either product substitution or green chemistry, in parallel to implementing planned or already required stormwater controls that intercept these pollutants once in stormwater, are the more sustainable approaches that are embraced by Bay Area stormwater agencies. For emerging contaminants, the responsibility for protecting surface waters from stormwater-borne pollutants must extend beyond the stormwater agencies themselves, and the management approach should include the more holistic evaluation of chemicals and products before they are allowed for use by federal (or state) agencies.

Lastly, rainfall patterns are changing in the Bay Area as a result of climate change. Larger and more intense storm events will likely increase the volume and intensity of runoff generated, which in turn may exacerbate existing, or introduce new, pollution problems to local creeks, rivers, and the Bay. More intense downpours can overwhelm the design capacity of

MS4s and lead to localized flooding and greater levels of pollutants in stormwater. On the other side of climate change, more frequent and extended drought conditions, which can lower the extent and duration of baseflow in our local creeks and rivers, are potentially impacting local species that rely on these freshwater ecosystems and drinking water supplies that rely on local rainfall. Changing hydrologic patterns and rising sea level pose their own set of issues, which underscores the need for dedicated federal and/or state funding to improve stormwater infrastructure and management.

Amid the fiscal challenges and impacts of new and emerging contaminants and climate change, diverse and integrated approaches to stormwater infrastructure design and management have emerged in the Bay Area. Traditional stormwater designs that rely on centralized "grey" infrastructure such as pipes, channels, and gutters which conveyed water quickly from urban streets, are being replaced by nature-based regional and distributed designs for LID and GSI. These approaches offer valuable opportunities to connect stormwater management goals with other planning sectors. Well-designed distributed GSI can support urban and water planning needs such as street beautification, multi-modal transit, transit-oriented development, water conservation, and groundwater recharge. Ongoing work at the state, regional, and local levels to integrate infrastructure planning for these various benefits should remain a priority at all scales and importantly from a coordination, partnership, and funding perspective. Coupled with an ongoing and enhanced focus on identifying and acting on new and emerging contaminants, the future of stormwater management over the next 50 years is a bright star on the constellation of water quality improvement, but must continue to be elevated with adequate funding and political support to ensure stormwater agencies (building on the knowledge gained through the RMP) can continue leading the charge to protect and restore the quality of surface waters in the Bay Area. §

Storm drain, Yerba Buena Island ▶
(photograph by Shira Bezalel).

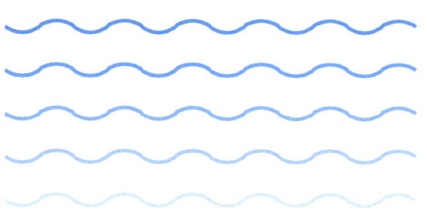

Dredgers

Josh Gravenmier · Arcadis
Bridgette DeShields · Integral Consulting Inc.
Maureen Dunn · Chevron
John Coleman · Bay Planning Coalition
Cameron Carr · Bay Planning Coalition

History and Need for Dredging

San Francisco Bay has been dredged for over 150 years (Dow 1973; USACE 1975; USACE et al. 1998, 2001) with the first legislation issued in 1863 by Governor Leland Stanford to authorize dredging to maintain the waters alongside docks, piers, and wharves in San Francisco and allow for further waterfront construction. The main ship channel in the Bay was first dredged in 1922 (USACE 1975), with multiple deepening projects continuing through the 1970s. Other port and harbor areas were created starting in 1920 and deepened, including the major deepening project at the Port of Oakland in 2009. Much of this dredged material was placed in the Bay, some of it serving as fill to create additional land mass with other material simply side-cast to adjacent areas. Beginning in the early 1970s, efforts increased to manage dredged material disposal, but disposal was limited to a few state and federally designated sites, with most material taken to a site near Alcatraz (USACE et al. 2001). From an average annual dredging of 7 million cubic yards (cy) by the US Army Corps of Engineers (Corps) alone in the 1970s and an additional 3-6 million cy dredged by others, total maintenance volumes decreased to approximately 3 million cy by the 1990s.

Evolution of Regulations

With few existing regulations, the establishment of the San Francisco Bay Conservation and Development Commission in 1969 by Governor Ronald Reagan (Dow 1973), provided a responsibility to regulate all filling and dredging in the Bay. In 1978, Public Law 95-269 further changed dredging operations in the Bay, as the Corps ceased being solely dredgers and became managers of dredging (Newton 1990).

The Federal Water Pollution Control Act, which, with amendments, became known as the "Clean Water Act" (CWA) in 1972 included provisions for dredged material in Sections 401 and 404. At the same time, the Marine Protection, Research, and Sanctuaries Act of 1972 required any proposed dumping of dredged material into ocean waters to be evaluated through the use of criteria published by the US Environmental Protection Agency (USEPA) (40 CFR 220-228), resulting in the creation of the "Green Book" (USEPA/USACE 1991). While early testing was rudimentary, suitability evaluations of dredged materials for disposal evolved, with an Inland Testing Manual (USEPA/USACE 1998) to further regulate "inland waters, near coastal waters, and surrounding environs." Building on these initial regulations, the current framework for the suitability determination process is very detailed and stringent, ensuring that any dredged sediment placed in the aquatic environment is considered "suitable."

The Dredged Material Management Office (DMMO), established in 1995, is an interagency virtual office with a mission to increase efficiency and coordination between the member agencies and foster a comprehensive and consolidated approach to handling dredged material management issues.

From a Waste to a Resource

For many years following the passage of regulations on dredging and disposal, dredged material was referred to as "spoils," portraying the material in a negative light and as material that could negatively impact water quality. However, as early as the 1970s, beneficial reuse of dredged material began being explored. Now clean and even mildly contaminated sediment is considered a valuable commodity.

With expanded regulations and increased value of dredged material usage realized, the Long-term Management Strategy for San Francisco Bay (LTMS; USACE et al. 1998, 2001) was initiated due to environmental concerns surrounding dredged material disposal and sought to limit in-Bay disposal to sustainable levels.

Economics and Necessity of Dredging

Maintaining navigation channels in the naturally shallow Bay is vital for our economy. Sufficient depths in navigational channels allow commercial industry vessels to operate at full capacity, reducing the need for lightering ships which can lead to increased emissions and potential for spills due to the increased numbers of ships. Examples of the importance of dredged shipping channels include the following.

- The total economic value of the marine cargo and vessel activity at the Port of Oakland, including the revenue and value-added at each stage of moving an export to the Port or import from the marine terminals, is estimated at $60 billion per year.

- The Bay Area manages 44% of the oil refining capacity in the state, with total annual state and local tax revenue from oil refineries estimated at $3.4 billion.

- The total number of jobs in the Bay Area supported by the oil and gas industry is estimated at 81,510 people. The Port of Oakland and its partners provide 84,144 jobs in the region.

Future Sediment Management

As ships continue to move in the Bay and potentially grow larger, dredging and possibly additional deepening will be necessary, but the shoreline around the Bay is at risk to sea level rise (SLR) impacts (both communities and sensitive infrastructure). The LTMS has been a fixture in the Bay since the year 2000, but what does the future look like?

Sea Level Rise: Effective Reuse of Dredged Material Is More Important Than Ever

A change in sediment regime, climate change, SLR, and other landscape drivers, without sediment augmentation, could cause the Bay's tidal marshes and flats to convert to a different habitat type, due to a lack of natural sediment supply to allow elevation gains to keep pace with SLR. According to the SFEI report "Sediment for Survival: A Strategy for the Resilience of Bay Wetlands in the Lower San Francisco Estuary," the volume of sediment needed for tidal marshes and tidal flats by the year 2100 is approximately 450 million cy, of which only about 30% will be supplied by current landscape and management approaches. Without additional sediment supplies, the Bay shoreline and associated communities risk inundation.

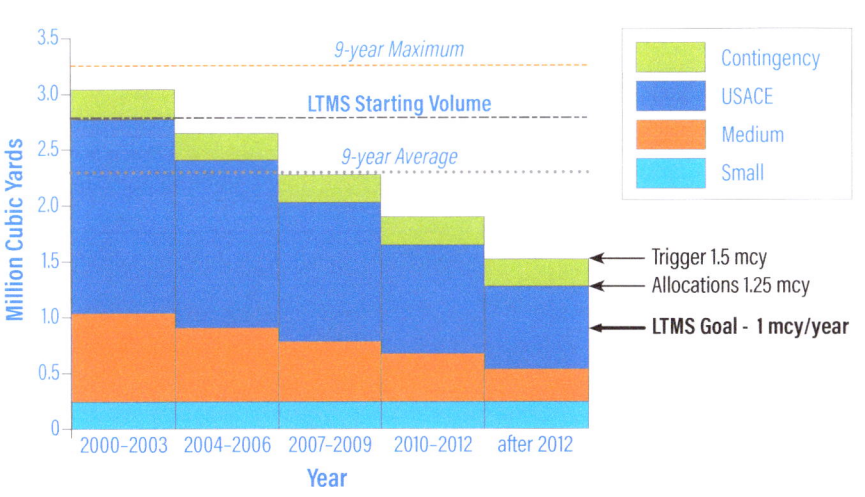

In-Bay Transition Allocations

Strategic Placement: Using the Power of Nature

Tidal marshes and flat areas protect us from SLR, king tides, and storms. Very few permitted restoration sites exist in the Bay, and they are only at specific locations and have limited capacity. The expansion of the Hamilton Wetland Restoration Project to include the Bel Marin Keys Unit V parcel would beneficially re-use 24.4 million cy of dredge material for habitat restoration, but additional sites are needed to accommodate the volume of sediment necessary to adapt to SLR. One option being considered is strategic placement of dredged materials in nearshore environments, using natural processes to bring the material onshore. Legislation from 2016 required the Corps to establish 10 pilot programs for the beneficial reuse of dredged material. The San Francisco District was funded for one pilot program and is currently evaluating a strategic shallow water placement of sediment to see if it would be a cost-effective method to create resilience.

Reuse Site Permit Streamlining: BRRIT Is Underway

The Bay Restoration Regulatory Integration Team (BRRIT) was formed in 2019 to improve the permitting process for multi-benefit habitat restoration projects and associated flood management and public access infrastructure in the Bay and its shoreline. The BRRIT began permitting projects and is showing promise for improving multi-benefit wetland restoration projects in the Bay as it brings together the six state and federal regulatory agencies.

New Cost-share Opportunities: Measure AA

Measure AA was approved in 2016 to raise $500 million via a 20-year parcel tax in the nine Bay Area counties, focused on building up the Bay's defense against SLR by restoring marshes for both habitat creation and flood protection, as well as improving public access. To date, almost $117 million in Measure AA funding has been authorized for 28 projects, restoring 6,402 acres to tidal marsh, tidal flat, and shallow bay areas. Funds can be used for dredging and beneficial reuse projects, but there is clearly not enough funding to meet current or future needs.

Regulations and Science: Updates Needed

The LTMS has been successful in implementing its desired goals and continues to have applicability into the future. Fundamentally, the LTMS goal is to minimize the cumulative environmental impacts and to maximize cumulative environmental benefits to the region as a whole. Similarly, the LTMS seeks to manage dredged material as a valuable resource for long-term benefits, as opposed to viewing it as a waste to be disposed of as inexpensively as possible in the short term. Even though the LTMS emphasized a balance between ocean disposal and beneficial reuse, with limited in-Bay disposal, it did not consider SLR or declining sediment loads to the Bay, as it was based on information from the late 1980s.

Available dredged material could be better utilized considering current and future conditions. We have to do things differently to bridge the anticipated gap in sediment volume needed to achieve resilience by the year 2100. Maximizing in-Bay sediment placement in appropriate near-shore locations and utilizing natural processes would be one helpful lever to pull. Another would be identifying additional funding streams (like Measure AA) to augment the costs to beneficially reuse dredge material without reducing the funding available to maintain navigation channels. Developing beneficial reuse locations throughout the Bay would directly increase resilience throughout the region. The BRRIT shows promise for improving the permitting process for multi-benefit wetland restoration projects, but is still constrained. Only by increasing state and federal funding, taking a regional perspective, and streamlining the permitting process will we enable dredged sediment to be utilized to the maximum extent possible to help mitigate SLR impacts and facilitate the movement of goods with fully maintained navigation channels. The RMP is providing valuable data, evaluations, and modeling that will assist in these efforts. §

Dredged Sediment Placement Locations*

* Sites located within LTMS Program Area, as of May 2020. Does not represent all sites where beneficial reuse is possible or has occurred.

** Defined as being available to receive dredged material within the next three years (Near Term Sites) or more than three years (Long Term Sites)

Source: 2021 Dredgers Handbook (https://www.spn.usace.army.mil/Portals/68/docs/Dredging/guidance/LTMS_DredgersHandbook_ADA_final_020221.ver=6yYVtXfmDBkYJpSjhruw9g%3d%3d)

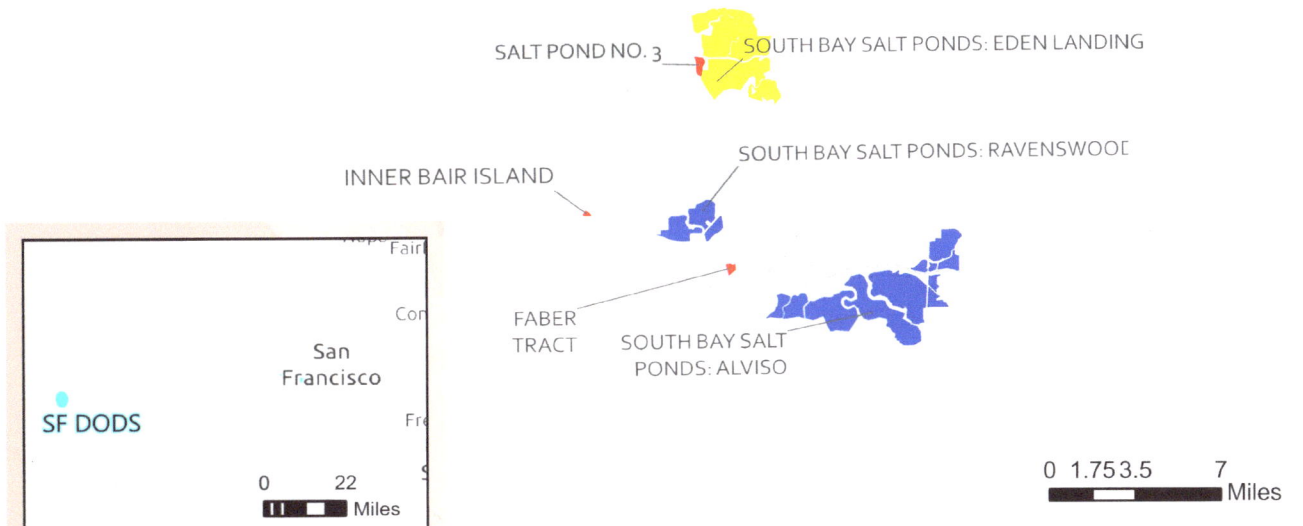

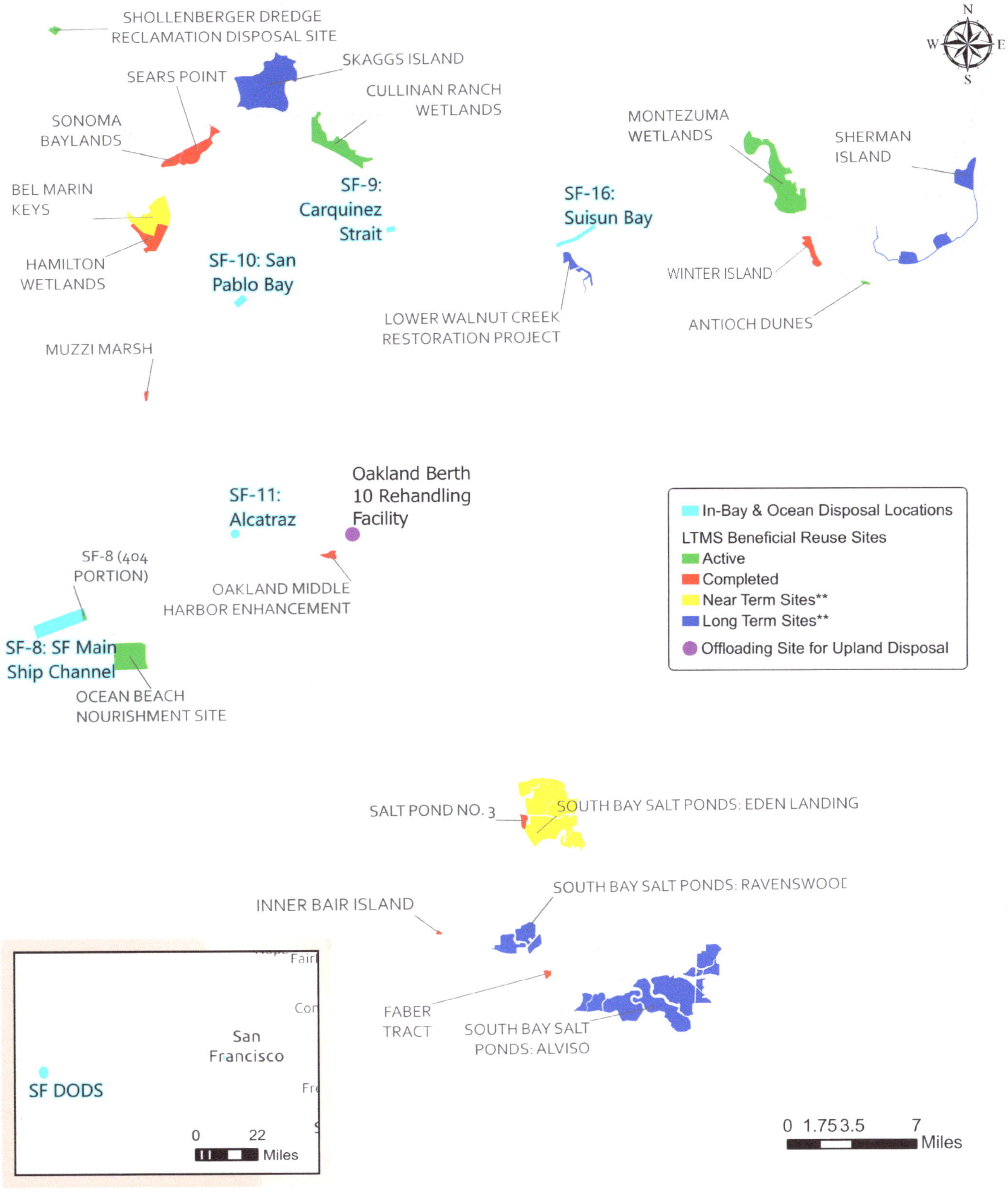

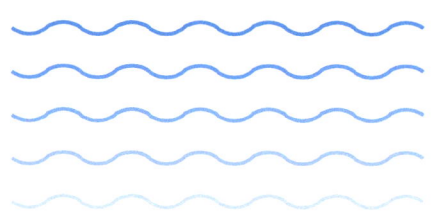

US Army Corps of Engineers

Tessa Beach

Although promulgated a half-century ago, the Clean Water Act of 1972 as amended (CWA) remains the seminal water quality protection statute in the United States today. The United States Army Corps of Engineers (USACE) is responsible for administering section 404 of the Act in order to "restore and maintain the chemical, physical, and biological integrity of waters of the United States through the control of discharges of dredged or fill material" (40 C.F.R. § 230.1). Fundamental to the protections afforded under the section 404 guidelines is the requirement that a discharge of dredge or fill material to waters of the United States, including wetlands, be prohibited if it would contribute to violations of State water quality standards (§ 230.10(b)), cause significant degradation of the nation's waters (§ 230.10(c)); or if a practicable alternative exists that is less damaging to the aquatic environment. Since the passage of the CWA 50 years ago, USACE — through its regulatory mission — has played a vital role in delivering water quality protection on a national scale by issuing individual and general permit decisions about discharges into waters of the United States and enforcing the provisions of such permits.

In the San Francisco Bay (Bay) region, USACE also has a primary mission to maintain safe and efficient navigation systems. Some of the most notable and innovative successes in managing Bay water quality to date have evolved at the nexus of these two mission areas. The foremost example is the establishment of the Dredge Material Management Office for the San Francisco Bay Region (DMMO) in the mid-1990s. As part of the Long Term Management Strategy for the Placement of Dredged Material in the San Francisco Bay (LTMS), the multi-agency DMMO established a comprehensive, coordinated, and efficient permitting approach to promote economically and environmentally sound placement of dredged sediment in the Bay region while operating within the existing laws, policies, and regulations governing each partner agency. The DMMO, chaired by USACE San Francisco District (District) since its inception, serves an important role in managing Bay water quality through review of sediment quality sampling plans, analysis of sediment quality results, and determination of sediment suitability for in-Bay, ocean, and/or upland placement. Since the year 2000, the DMMO has provided hundreds of suitability determinations for dredging projects in the Bay, which in total cover nearly 65 million cubic yards (MCY) of material.[1] Moreover, the DMMO provides a clearinghouse of sediment chemical and biological quality data through its web database that can be queried regionally and is used to inform water quality management studies by, for example, scientists at the San Francisco Estuary Institute.

While USACE San Francisco District plays an important role in Bay water quality management through Section 404 permitting and the DMMO, the District also carries out a debris removal mission in the Bay and executes a robust dredging program to operate and maintain federal navigation channels, both of

---

[1] USACE 2022. Dredged Material Management Office (DMMO) Dredging and Placement of Dredged Material in San Francisco Bay January-December 2021 Report. https://www.spn.usace.army.mil/Portals/68/docs/Dredging/Annual%20Reports/2021%20DMMO%20Annual%20Report_Final.pdf?ver=lgVsonooeCN7yMczI-2gqg%3d%3d

which have contributed noteworthy water quality benefits over the last 50 years. Since first charged with the mission in the 1940s, the District has continuously removed debris such as pilings and sunken vessels from Bay waters. Today the District operates two debris hazard collection boats that patrol Bay waters and remove approximately 90 tons of debris a month[2] that might otherwise pollute the Bay, to the benefit of both navigation safety and water quality. The District also serves as the largest dredger in the Bay (by annual volume) through its operation and maintenance (O&M) of the Bay's twelve federal channels as well as the San Francisco Main Ship Channel on varying annual to semi-annual cycles. The sediment dredged from these channels can be a valuable resource to restore wetlands that provide numerous environmental benefits including those to water quality, for example via the filtration of constituents that would otherwise enter the Bay. Overall, between 2000 and 2021, the District dredged 36 MCY of material from in-Bay federal channels, of which approximately 10 MCY was beneficially used for wetland restoration. A significant example of success in this regard is the Hamilton Wetlands Restoration Project, a partnership between the USACE and the California State Coastal Conservancy (SCC), which beneficially reused dredge material to restore 648 acres of wetland habitat.

Looking ahead to the next 50 years, the effects of climate change and sea level rise will increasingly threaten aquatic ecosystems and the physical and social infrastructure of the Bay. The need for regional resiliency and adaptation to these threats will drive further demand for the beneficial reuse of sediment to support wetlands, tidal marshes, and mudflats. Moreover, sustaining existing and restoring additional Baylands into the future will require an increasing amount of sediment as sea level continues to rise throughout the 21st century. Dredged material will be a crucial resource to address these needs. Yet scientific, logistical, economic, and regulatory obstacles associated with beneficial reuse of dredged sediment remain. USACE implements its O&M dredging program within the context of the Federal Standard. This means that USACE regulates dredge material "…to assure that dredged material disposal occurs in the least costly, environmentally acceptable manner, consistent with engineering requirements established for the project" (33 C.F.R. § 336.1). In San Francisco Bay, where direct upland beneficial reuse of sediment is regularly the most expensive placement option, reducing and/or sharing the incremental cost of beneficial use is particularly critical to being able to accomplish more of it.

This is not an insurmountable challenge! For instance, in recent years, the SCC formed a successful partnership with the San Francisco District to fund the incremental cost of taking material from the Redwood City federal channel to upland beneficial use. Cost reductions, on the other hand, may be achieved by creating more available capacity for beneficial use through the establishment of new restoration locations around the Bay and innovative placement practices. To that end, the District and SCC, in collaboration with the Bay Conservation and Development Commission and the San Francisco Regional Water Board, are piloting an innovative shallow-water strategic placement pilot project to evaluate the feasibility of depositing dredged material from a federal channel in the shallow margins of the Bay and relying on natural hydrodynamic processes to beneficially deliver that material to nearby intertidal Baylands. Similarly, the LTMS agencies are pursuing a pilot that would increase beneficial reuse by diverting approximately 50% of the material from the Oakland Harbor Federal Channel to an upland beneficial use site and permitting the other approximately 50% to be placed in-Bay to offset the higher cost of the beneficial use. Finally, USACE and the SCC are seeking to partner on the Bel Marin Keys wetland restoration project, which would bring online a new Bay site with significant additional capacity to accept dredged material for restoration purposes. These approaches take true collaboration and creativity, and the Bay region is fortunate to have many agencies, organizations, and communities committed to working together to increase beneficial use of dredged material and restore Baylands. Through such collaboration and innovation, we can overcome obstacles to more widespread beneficial use to ensure the resiliency of the Bay, and its aquatic resources, in the face of climate change. §

---

[2] USACE 2022. San Francisco Bay Hazard Removal Program. Accessed June 29, 2022 at https://www.spn.usace.army.mil/Missions/Hazard-Removal-SF-Bay/

Photograph · Shira Bezalel, August 2022

2022

Remnants of the Dumping Ground Era

A portion of Point Isabel in Richmond, locally known as TEPCO beach, is covered in thousands of pieces of broken ceramic dishware. The Technical Porcelain and Chinaware Company (TEPCO) was founded in 1918 and shut down its operation in 1968. During that span, TEPCO became a major West Coast producer of decorative, durable hotel and restaurant ware.

TEPCO supplied not only the hospitality industry, but had contracts with the US government to produce ware for the Navy, Army and Veterans Administration, and was, for years, El Cerrito's largest employer. Hundreds of people worked at the factory, making tens of thousands of pieces of pottery every day in a wide variety of designs. TEPCO dishes were everywhere. Local Bay Area restaurants like Louie's Restaurant

Club, Doggie Diner, and Spenger's Fish Grotto had custom designed TEPCO plates. TEPCO dishes were even used at the Kaiser shipyards. The US Army and Navy used full sets of TEPCO in their mess halls and on their ships. TEPCO dishes are now sought after by avid collectors.

As was common practice back then, TEPCO disposed of its chipped and damaged dishes in the Bay, on the shore of Point Isabel. The porcelain fragments are still readily apparent over 50 years after the dumping ended — a visible reminder of the Bay's dumping ground era.

Another, less innocuous form of dumping also occurred at Point Isabel. A cove on North Point Isabel was once piled high with industrial batteries that leached lead and zinc into the Bay. In the mid-1980s, contaminated debris was removed and North Point Isabel was sealed with a clay cap. In 2002, it became part of McLaughlin Eastshore State Park, which is managed by the East Bay Regional Park District as an extension of the Point Isabel Regional Shoreline. Recent soil testing detected lead in specific, localized areas around the perimeter of the north side of the park. The areas are currently fenced off from all public and pet access. The Park District, under oversight of the San Francisco Bay Regional Water Board, is developing a plan to restore these areas.

Photos of the Tepco factory, 1955.

Top left: Tepco factory in El Cerrito, built in 1947 to replace earlier buildings that were damaged by fires.

Top right: Tepco employees sanding plates.

Bottom: Back view of the factory, with piles of broken and discarded china (likely taken to Point Isabel).

Photos courtesy of r_leontiev, Flickr, Creative Commons 2.0.

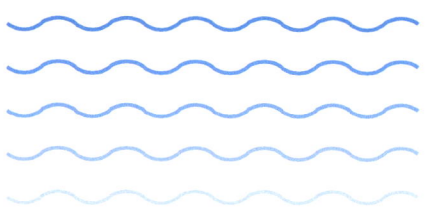

Baykeeper

Ian Wren

Clean Water Act: May You Be Forever Strong!

In the early 70s, when the Clean Water Act passed Congress, the San Francisco Bay Area was already at the center of an emerging ecology ethos that would eventually spread across the nation.

Perhaps the residents of the Bay Area were shocked into this new perspective by the audacity of an aggressive Bay-fill plan put forth by the Army Corps of Engineers. It would have reduced the Bay we know today into little more than a rivulet, surrounded by sprawling new communities built on top of landfill that was once Richardson Bay, Suisun Marsh, the lowlands of the South Bay, and the heart of San Francisco Bay itself.

The value of any real estate along the Bay was judged solely in terms of its availability for commercial development. Little if any thought was given to preserving natural open spaces, or to acknowledging that the Bay Area's residents and visitors might value the region's unique beauty.

In 1965, Bay Area elected officials in Sacramento were able to pass a moratorium on Bay fill. In 1969, the legislature formed the Bay Conservation and Development Commission to protect and enhance the Bay for future generations.

But even so, the Bay was ringed by industrial pollution source points, and it was commonplace for municipal wastewater treatment plants to flush raw sewage directly into the Bay. Pollution was rampant, and the regulatory agencies, such as they were, didn't have the legal tools necessary to stop it.

The Clean Water Act changed all of this. This new law gave state and federal agencies the authority to hold polluters accountable—and equally important, it provided funding for pollution prevention programs. There were now legal controls to manage significant pollution dischargers like wastewater treatment facilities and refineries.

The fundamental genius of the law—indeed, its moral core—is that it recognizes that every living being has the right to a healthy environment. The Clean Water Act states unequivocally that "any person" has the legal right to make sure that the provisions of the law are being upheld and can sue to enforce federal or state-issued water quality standards. That gives environmental advocates such as Baykeeper the legal footing we need to defend access to clean and healthy waters for all.

The law would continue to get stronger. Amendments adopted in the late 1970s gave the Environmental Protection Agency the authority to manage wetlands, which is crucially important to the health of San Francisco Bay. In the 1980s, Congress further amended the law to enact significant additional controls on industrial and municipal stormwater pollution.

Despite these developments, the Clean Water Act can be improved. The Bay and nearby creeks remain impaired by numerous pollutants. Many of the Bay Area's established toxic hot spots have not been cleaned up. Regulatory agencies have known for decades that sites like Richmond's Selby Slag and AstraZeneca are dangerously polluted and pose significant threats to the Bay's water quality and the health of residents in surrounding neighborhoods. These legacy sites often have complicated histories of ownership, so it's not always clear who to hold responsible for remediation.

Congress could strengthen the Clean Water Act to require pollution management at the watershed scale and provide the necessary funding mechanisms to allow the cleanup of hotspots at a regional scale.

Additionally, many pollutants have only recently been identified as dangerous. These next-generation pollutants include a variety of compounds such as antibiotics, drugs, steroids, endocrine disruptors, hormones, industrial additives, and chemicals, as well as microplastics and materials associated with the high-tech industry.

These pollutants can wind up in municipal and industrial wastewater before entering the Bay and its watershed. Many of these emerging pollutants remain unregulated under state and national legislation, which only increases their risk to environmental and human health. The Clean Water Act would certainly be strengthened and more relevant to today's realities if it included these emerging pollutants and provided a means to control them.

Yet the Clean Water Act remains the most critical piece of legislation that has protected America's waterways from pollution, and the most useful. That's precisely why industry groups and development interests relentlessly lobby against it—sadly, they are making gains.

The conservative super-majority on the Supreme Court will have a chance to solidify these gains in an upcoming case on its docket in fall 2022, Sackett v. EPA. In that case, the parties bringing the lawsuit and the industry lobbyists supporting them are urging a substantial narrowing of waters of the United States, attempting to limit the law's protection to only traditionally navigable waters and those with a surface connection to them. Never mind that bodies of water can be connected by underground aquifers, or by the tides, or that science demonstrates that wetlands significantly affect the chemical, physical, and biological health of adjacent waters. And never mind that the Clean Water Act mandates a much broader scope of protection.

There would be consequences for San Francisco Bay. For instance, corporate giant Cargill wanted to pave over the South Bay salt ponds for commercial development, and Trump's EPA tried to help them do it. In the ensuing case, Baykeeper v. EPA, the agency's attorneys argued that the ponds were land, not water, and that they had no real connection to the Bay. However, the court determined that salt ponds in Redwood City are water, are clearly connected to the Bay, and held that the EPA had misapplied the Clean Water Act. This now leaves the salt ponds available to be restored into wetlands that can help buffer the South Bay from storms and climate-driven sea level rise. But if the Supreme Court rules as requested in Sackett v. EPA, there's no guarantee that common sense would prevail in the future.

Such an interpretation would be short-sighted, but that's not the half of it. Earlier this year, the Supreme Court reinstated the Trump EPA's rule that defined protected waters in a way that strips protections from seasonal streams and isolated waters such as vernal pools. This interpretation could strip federal protection from about 70 percent of the creeks that flow to San Francisco Bay. That percentage would be even greater in the more arid regions of the West.

The Biden administration is taking action to put its own science-based regulations in place of the so-called Trump rule. It also announced a new set of rules to strengthen the authority of states, territories, and Tribes to protect their vital water resources, thereby restoring long-held rights under the Clean Water Act that were severely curtailed under Trump.

Regardless of what happens, the Clean Water Act will continue to be our strongest tool to hold polluters accountable, and we will still have the basic human right to take legal action to protect our waters. Happy birthday, Clean Water Act. May you be forever strong! §

2022 WATER QUALITY PARAMETER SUMMARIES

BACTERIA

Origins of the Problem

The Gold Rush of 1849 initiated rapid population growth in the Bay Area and a 100-year period of ever-increasing discharge of untreated sewage to the Bay. Pathogenic organisms found in fecal waste from humans and other warm-blooded animals (including pets, horses, livestock, and wildlife) can cause infections and illness in people who come into contact with contaminated waters or consume contaminated shellfish.

In the early 1900s, government health authorities understood the hazards of swimming in sewage-contaminated Bay water (pages 6-7). Health risks associated with shellfish consumption were also understood: by 1940 a lucrative oyster fishery was closed as a result of bacterial contamination.

The first sewage treatment facilities were constructed in the 1950s and commonly used chlorine as a disinfectant to reduce pathogens. However, these facilities only provided primary treatment (settling and mechanical filtering), and chlorination is less effective on partially-treated wastewater. Until the widespread adoption of secondary treatment of wastewater (reduction of organic matter through bacterial metabolism) in the 1970s, spurred by the Clean Water Act (CWA), the constant flow of untreated or minimally treated sewage was the major source of the overabundance of pathogens present in the Bay.

While improved treatment and relocation of outfalls to deep water resulted in vastly enhanced water quality, pathogens remain a problem in the Bay today. Present concern is focused on urban beaches, where the major sources and pathways of bacteria include overflows from sewage collection systems, urban stormwater runoff, pets, vessels, and wildlife.

History of Monitoring

Early monitoring in the Bay had a greater focus on assessing the safety of shellfish consumption. A 1972 USEPA study reported that more than 50% of the waters directly over known Bay shellfish beds contained coliform bacteria densities in excess of state and federal criteria for shellfish-growing waters. The wastewater treatment upgrades of the 1970s led to rapid improvements. Total coliform counts, which averaged 800 organisms per 100 ml in the South Bay in 1964, had declined to an average of 4 organisms per 100 ml in 1977. By 1982 public harvesting of shellfish in San Mateo County was approved for the first time in 50 years. By 1987 the Water Board concluded that swimming was safe in most areas of the Bay during summer.

In the past two decades, monitoring has focused more on assessing conditions for swimming and other forms of contact with the water. Beginning in 1999, Assembly Bill 411 required bacterial testing between April 1 and October 31 of waters adjacent to major public beaches near storm drains. As a result, county public health and other agencies routinely monitor fecal indicator bacteria (FIB) concentrations at 24 Bay beaches and provide warnings to the public when concentrations exceed the standards.

History of Advisories and Regulation

Bacteriological standards to protect swimmers and shellfish consumers existed for decades before the CWA, supporting swimming warnings and shellfish closures like those mentioned above. The CWA era of regulation began with the adoption of the 1975 Basin Plan, which established numerical water quality objectives for bacteria and other constituents. The Basin Plan called for discharge limits for coliform bacteria and prohibitions of wastewater discharges that did not receive a minimum initial dilution of 10 to 1 or to confined water areas or their immediate tributaries. These regulations guided the wave of treatment plant upgrades in the 1970s.

In spite of the treatment plant upgrades, exceedances of bacteria objectives still occur due to other sources. At present, eight Bay beaches are on the 303(d) List of impaired water bodies because fecal indicator bacteria exceed standards. The CWA requires that TMDL control plans be adopted for water bodies on the 303(d) List. In 2017 a TMDL was approved to address bacteria at six beaches (China Camp, McNears Beach, Crissy Field, Aquatic Park, Candlestick Point, and Marina Lagoon). Another TMDL for five more Bay beaches (Erckenbrack Park, Gull Park, Marlin Park, Kiteboard Beach, and Oyster Point Marina) is in development.

Current Status and Long-term Outlook

When it's not raining, Bay beaches are generally safe places to enjoy recreation in the water. In wet weather, however, bacteria concentrations often exceed water quality standards and increase the risk of illness or infection.

Using the county beach monitoring data, Heal the Bay, a Santa Monica-based non-profit, provides Beach Report Cards for over 400 California bathing beaches in as a guide for beach users. The Bay-wide average summer grade for 24 monitored beaches in 2021 was an A (GPA of 4.0). This was the highest summer grade observed for the period of record since 2003. During wet weather, bacteria concentrations are considerably higher due to stormwater runoff and sewer overflows. The overall average GPA for the 24 beaches in wet weather was 1.4 (a D+).

Successful implementation of the TMDLs is expected to improve conditions at the beaches with higher fecal indicator bacteria concentrations. The beaches TMDL approved in 2017 incorporates management measures to reduce or eliminate waste discharges from sanitary sewer systems, stormwater runoff, vessels, pets, and controllable wildlife. However, extreme weather and flooding associated with climate change could exacerbate fecal contamination from sewer overflows and stormwater runoff and make progress more difficult. §

Bay-Wide Average Summer Beach Report Card Grades

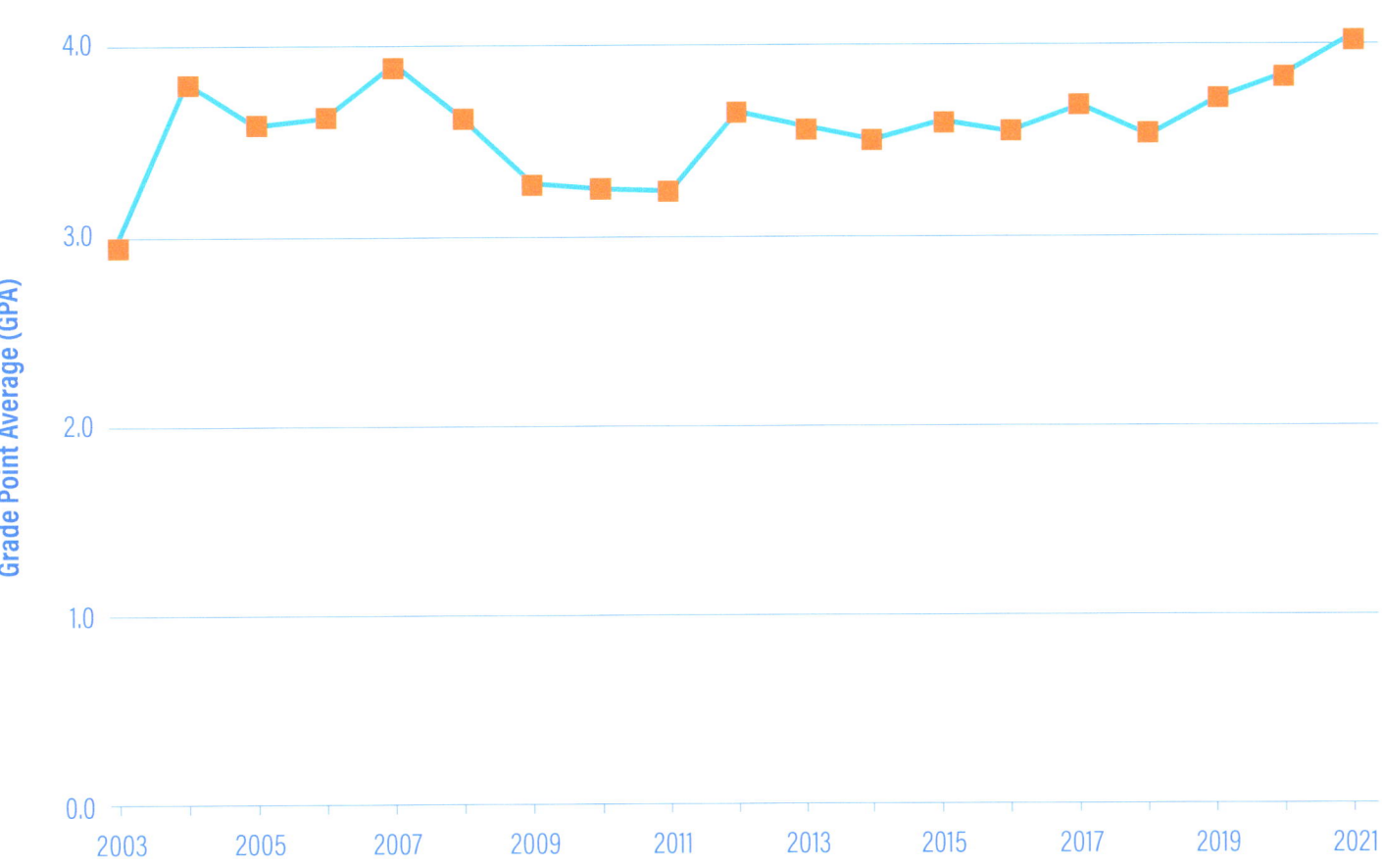

FOOTNOTE: Data from Heal the Bay (2022) and earlier Heal the Bay reports.

WATER QUALITY PARAMETER SUMMARIES

ORGANIC WASTE

Origins of the Problem

Prior to the Clean Water Act of 1972 (CWA), the discharge of oxygen-depleting organic waste from municipal and industrial facilities was perhaps the Bay's biggest water quality problem. Dissolved oxygen (DO) is vital to aquatic organisms. When environmental microbes metabolize organic waste, such as the solid material in untreated sewage, they deplete oxygen from the water column. When oxygen levels become too low, fish and other organisms can suffocate and die.

The population of the Bay Area grew from 2,000 in 1850 to 2.7 million in 1950, and ever-increasing quantities of untreated sewage were discharged into the Bay. By the 1950s, many Bay Area communities had built primary treatment plants, which removed material that could be screened or would either float or readily settle out by gravity. This minimal level of treatment left large amounts of organic waste and other sewage pollutants flowing into the Bay, and the Bay suffered from low DO, frequent fish kills, foul odors, high concentrations of fecal bacteria, and other problems.

Secondary treatment removes 80% to 90% of oxygen-demanding organic waste through microbial metabolism. Just a few facilities, including San José-Santa Clara, Oro Loma, and Dublin-San Ramon, were providing secondary treatment by the late 1960s. Along with the NPDES Program, the CWA provided clear goals and over a billion dollars toward widespread adoption of secondary treatment in the Bay Area. By 1987, all municipal wastewater treatment plants discharging to the Bay were providing at least secondary treatment. By 1985, Bay Area municipal wastewater treatment plants had reduced biochemical oxygen demand (BOD) loading by 88% from the high levels recorded two decades earlier, while the service area population increased by 52% over the same period.

History of Monitoring

The first measurements of DO in the Bay were made in the late 1950s and they showed recurrent summer anoxia in the Lower South Bay (Cloern and Jassby 2012).

The US Geological Survey (USGS) has had a central role in monitoring DO and other basic water quality parameters in the Bay. In 1969, the USGS began a systematic program of regular monitoring throughout the Bay. This program has been sustained to this day, providing one of the longest records of water quality measurements in a North American estuary (Schraga and Cloern 2017). Monitoring of DO as part of this program began in 1971 and continued through 1978. Following a gap from 1979-1992, monitoring resumed in 1993. Monitoring of DO and other parameters on these USGS cruises since 1993 has been partially supported by the RMP.

Lower South Bay has historically been particularly impacted by organic waste because these waters are shallow and have relatively limited exchange with the rest of the Bay. The long-term time series for DO in Lower South Bay documents a dramatic improvement in Bay water quality as a result of the Clean Water Act (Cloern and Jassby 2012). Secondary treatment was fully implemented by all treatment plants discharging to lower South Bay by 1973. Summer anoxia was eliminated in the 1970s, but DO concentrations still sometimes fell below 5 mg/L, a common standard to protect marine fish. In the late 1970s, the three Lower South Bay treatment plants further contributed to the BOD load reductions in this region by installing nitrification and filtration. By 1985, dischargers to this region had decreased their BOD loading by 99% even though wastewater flows had more than doubled since 1955. The most recent part of this time series (1993-2011) shows the virtual elimination of DO concentrations below 5 mg/L.

More intensive monitoring of DO in recent years using a network of moored sensors has been conducted as part of the Nutrient Management Strategy (page 52) to evaluate oxygen depletion resulting from algal blooms.

History of Advisories and Regulation

Regulation of sewage discharge in its initial stages was principally a matter of managing organic waste. Little regulation or sewage treatment occurred before the 1950s. A 1949 report to the Legislature noted that the sewage treatment had not changed much in 100 years (SFBRWQCB 2000).

The Dickey Water Pollution Act of 1949 established the State and Regional Water Boards and gave the regional boards the ability to set discharge limits, but little enforcement authority. Under the Dickey Act, cities and industries implemented more wastewater treatment in the 1950s and 1960s, but it was not enough to keep up with population growth. The amount of BOD discharged to the Bay peaked in the mid-1960s.

The Porter-Cologne Act of 1969 gave the Regional Board authority to set standards, issue orders to implement those standards, and, most importantly, the ability to enforce its orders.

While some cities implemented secondary treatment under these earlier state laws, it was the CWA that drove general adoption of secondary treatment by requiring it and providing substantial funding for treatment plant construction and upgrades.

The first Basin Plan, adopted in 1975, included a number of important elements that helped to control oxygen depletion, including water quality objectives for DO and un-ionized ammonia, a requirement for 10:1 dilution, and tertiary treatment requirements for certain facilities.

Current Status and Long-term Outlook

Today, Bay Area wastewater treatment plants are effectively controlling the input of organic waste to the Bay. Current concern for oxygen levels is due to the potential for algal blooms fueled by the Bay's high nutrient concentrations (page 52). §

"Sewage from San José and other cities was discharged without treatment to San Francisco Bay. This practice has resulted not only in gross pollution of the receiving waters but has become a principal cause of a seasonal atmospheric condition manifested over a wide area by a sulfide odor, a tarnishing of household silver, and a blackening of painted surfaces."

— From a 1953 report prepared for the City of San José. Cited in SFBRWQCB (2000).

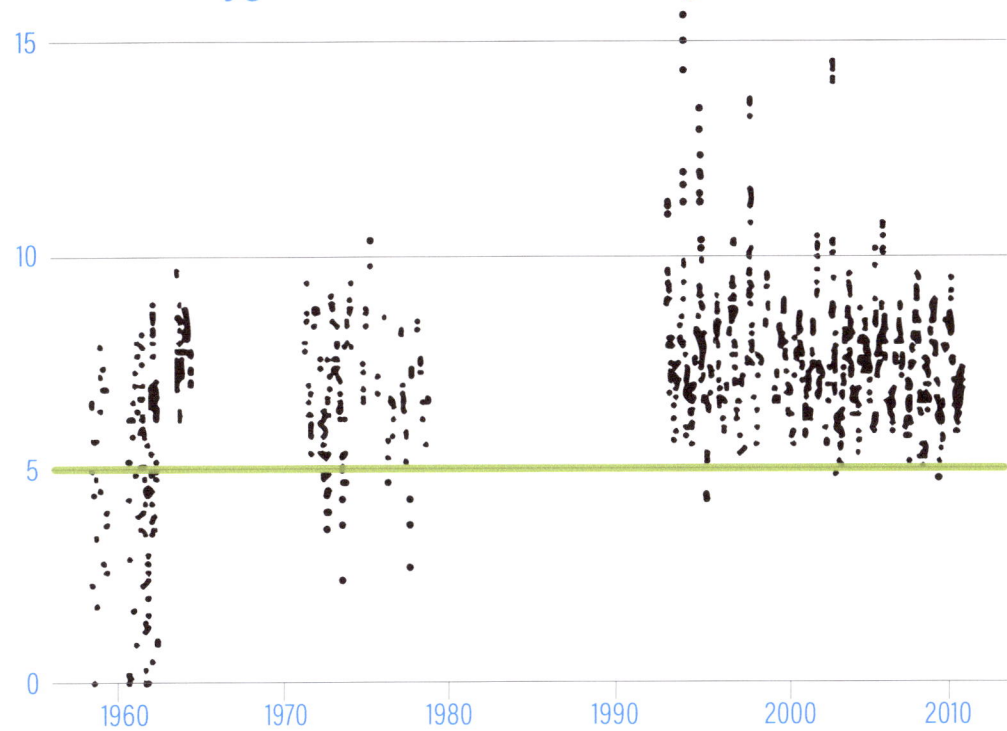

FOOTNOTE: The green line represents a common standard to protect marine fish sensitive to low oxygen. From Cloern and Jassby (2012).

BOD Load to the Bay (1955-1985)

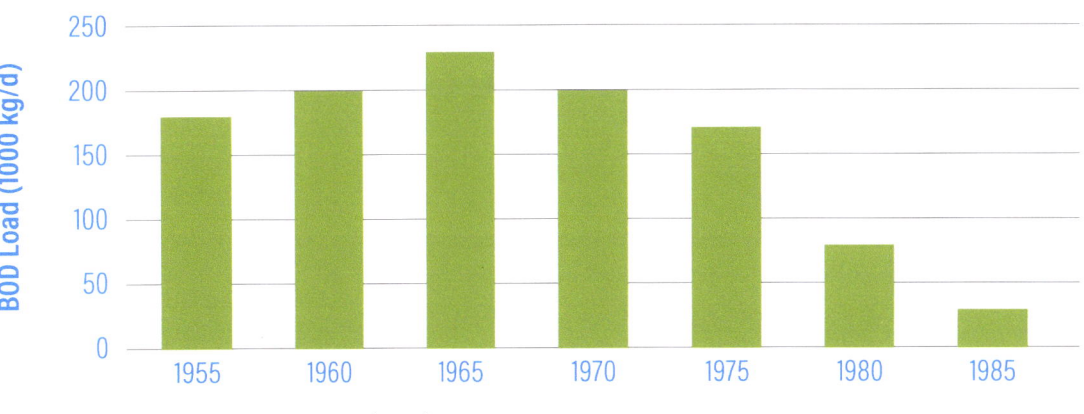

FOOTNOTE: From SFBRWQCB (2000).

WATER QUALITY PARAMETER SUMMARIES

NUTRIENTS

> The text and graphics for this section were developed prior to a major harmful algal bloom in the Bay in late August 2022, and do not address that event. SFEI and partner organizations conducted extensive monitoring during the bloom. The data generated will be assessed as part of a thorough analysis by the Nutrient Management Strategy (NMS) of the possible causes of the bloom. Updates will be provided via the NMS website.

Origins of the Problem

Nitrogen (N) and phosphorus (P) are natural and essential components of healthy estuarine ecosystems. Sufficient nutrient levels are needed to support the growth of phytoplankton (also referred to as algae) that in turn serves as the base of the food web. Excess nutrient loads resulting from anthropogenic activities can, however, lead to a range of adverse impacts, including excessive phytoplankton production, low dissolved oxygen, and harmful algal blooms (HABs).

San Francisco Bay receives elevated loads of N and P, and ranks among the most nutrient-enriched estuaries worldwide (Cloern et al. 2020; SFEI 2014a,b). Treated effluent discharged by the Bay Area's 37 publicly-owned treatment works (POTWs) accounts for 65% of annual-average nutrient loads Bay-wide, and more than 80% in some regions of the Bay (SFEI 2014a). Other anthropogenic sources include flows from the Delta and stormwater runoff from Bay Area watersheds. Despite its nutrient-enriched status, the Bay has historically been spared the impacts common to many other nutrient-enriched estuaries. The Bay's 'resistance' to elevated nutrients has been attributed to multiple factors, in particular high suspended sediment concentrations that limit light needed for phytoplankton growth, and strong tides that thoroughly mix the water column.

However, a growing body of evidence, assembled over the last 15 years, suggests that the Bay's resistance to elevated nutrients may be waning, including: increased phytoplankton biomass in deep subtidal regions of South Bay (Cloern et al. 2007) and Central Bay (SFEI 2022); frequent occurrences of harmful algae and their associated toxins (Sutula et al. 2017; Peacock et al. 2018); and low dissolved oxygen in some tidal slough habitats (SFEI 2015, 2018, 2021). To address these concerns, regulators and stakeholders initiated the San Francisco Bay Nutrient Management Strategy (NMS) in 2012.

History of Monitoring

The USGS began conducting water quality monitoring in the Bay in the late 1960s (Schraga and Cloern 2017), with biweekly to monthly data available along the deep channel since the early 1980s for a range of nutrient-related parameters. In 1993, the RMP began partnering with USGS on water quality monitoring by supporting a portion of the annual program budget. Data and interpretations from that on-going work have played a fundamental role in assessing water quality conditions and in shaping our understanding of the Bay's response to nutrients.

Nutrient-related monitoring has increased with the launch of the NMS. The NMS collaborates with USGS and UC Santa Cruz on water quality cruises along the Bay's deep channel, in particular through supporting nutrient measurements; phytoplankton taxonomy; measurements of harmful algal toxins; and the development of DNA-based techniques for detecting and quantifying harmful algae. In addition, over the last several years, SFEI has installed a network of moored sensors in South Bay and Lower South Bay that measure dissolved oxygen, chlorophyll-a, and other water quality parameters every 15 minutes. Some sensors overlap spatially with USGS deep-channel cruises, while others are measuring conditions in important regions not regularly sampled by long-term monitoring, e.g., South Bay's broad shoal and slough habitats in Lower South Bay. SFEI is also collaborating with the USGS Biogeochemistry group (California Water Science Center) to conduct high-speed biogeochemical mapping cruises targeting South Bay shoal habitats. The mooring and mapping work provide both additional spatial coverage and high-frequency data that are critical for understanding the dynamic processes affecting dissolved oxygen and nutrient cycling in this portion of the Bay. Over the last five years, SFEI and UC Santa Cruz have also been measuring algal toxins in native mussels collected every two weeks from docks around the edge of the Bay. The mussels serve as time-integrated samplers of algal toxins entering the food web.

Advisories and Regulation

Regulators and stakeholders initiated the NMS in 2012 to address concerns related to evidence of the Bay's shifting response to nutrients. One of the earliest NMS actions was to begin monthly monitoring of nutrient loads from all POTWs (prior to 2012, nutrient discharges to San Francisco Bay were not regulated, and monitoring of POTW nutrient loads had not been required). In 2014, the Water Board issued the first Bay-wide nutrient permit, a 5-year permit that called for POTWs to provide funding to support a NMS Science Program charged with developing the scientific foundation to support nutrient management decisions. The Water Board also convened a 15-person NMS Steering Committee, composed of representatives from stakeholder groups (regulators, dischargers, water purveyors, non-governmental organizations, and resource agencies), to oversee the NMS Science Program, including financial oversight and high-level input on programmatic priorities. SFEI serves as the NMS Science Program's technical lead, and SFEI staff work closely with regional collaborators to carry out NMS-sponsored field investigations, monitoring, and data interpretation. The first Bay-wide nutrient permit also required POTWs to investigate the opportunities for and costs of reducing nutrient loads.

In 2019 the Water Board issued the second 5-year Bay-wide nutrient permit. In addition to continued support for the NMS Science Program, the second permit required POTWs to evaluate opportunities to reduce nutrient discharges using "green" solutions, like natural systems (e.g., wetlands) and wastewater recycling — opportunities that can provide multiple benefits beyond nutrient removal.

Current Status and Long-term Outlook

Nutrient loads to San Francisco Bay were already elevated in the early 2000s. Between 2005 and 2018, N loads from the five largest POTWs increased by 35% (Figure), generally consistent with population increases over that time, since human waste is the source of most wastewater-derived N. Loads subsequently decreased substantially over the first 1.5 years of the COVID pandemic.

Phytoplankton biomass (measured as chlorophyll) and dissolved oxygen are two important nutrient-related indicators of water quality. In addition to documenting current conditions, NMS work is also focused on tracking changes in key condition-indicators over time.

Decades of phytoplankton biomass observations in deep subtidal regions of South Bay (1970s-1990s) had suggested that San Francisco Bay was less sensitive to nutrient enrichment than many other estuaries. That same time-series subsequently showed that summer-fall chlorophyll levels in South Bay had more than doubled between the early 1990s and 2005 (Cloern et al. 2007). The tracking of chlorophyll has continued, with data suggesting that Aug-Oct biomass levels in far South Bay have decreased substantially since 2008, returning close to early-1990s levels by 2019 (Figure). Unlike far South Bay, Aug-Oct chlorophyll levels in Central Bay doubled and have remained elevated through 2019. Understanding the underlying causes of these differences in behavior is an important focus of current work. While Aug-Oct phytoplankton biomass levels have exhibited substantial changes, initial analyses suggest that Aug-Oct dissolved oxygen levels have not undergone significant changes, and remained well-above the Basin Plan standard of 5 mg/L. Analysis of the long-term chlorophyll and dissolved oxygen data are continuing, including to understand the factors or processes contributing to changes over time.

The data discussed above were collected in the Bay's deep channel. NMS-supported monitoring work over the past several years indicates that water quality conditions in nearby habitats can differ substantially from deep-channel conditions. For example, some Lower South Bay sloughs have highly-elevated phytoplankton levels and dissolved oxygen concentrations frequently drop below 5 mg/L, due in part to high oxygen consumption rates. Work is underway targeting improved quantitative understanding of the processes leading to these conditions, and developing scientific guidance for identifying protective dissolved oxygen levels in these habitats. The data discussed above focus on deep subtidal habitats. NMS-supported monitoring work over the past several years indicates that water quality conditions can vary substantially between nearby habitats, including shoal versus channel in South Bay and slough versus open-Bay in Lower South Bay. §

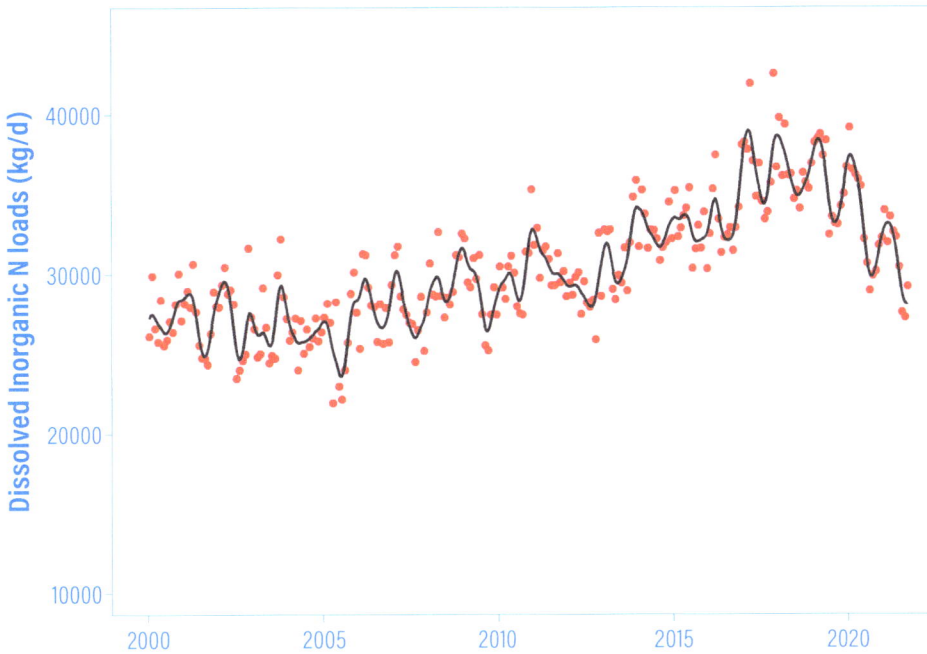

Dissolved Inorganic Nitrogen Loads from the Five Largest POTWs

FOOTNOTE: Dissolved inorganic N loads (kg/d) from the five largest POTWs discharging to the Bay (EBMUD, SFPUC, EBDA, San Jose, CCCSD). Black circles: summed monthly load estimates from the five POTWs (Data: SFEI 2014; BACWA/HDR 2022); red curve: GAM model fit (see Beck et al. 2021). SFEI, in preparation.

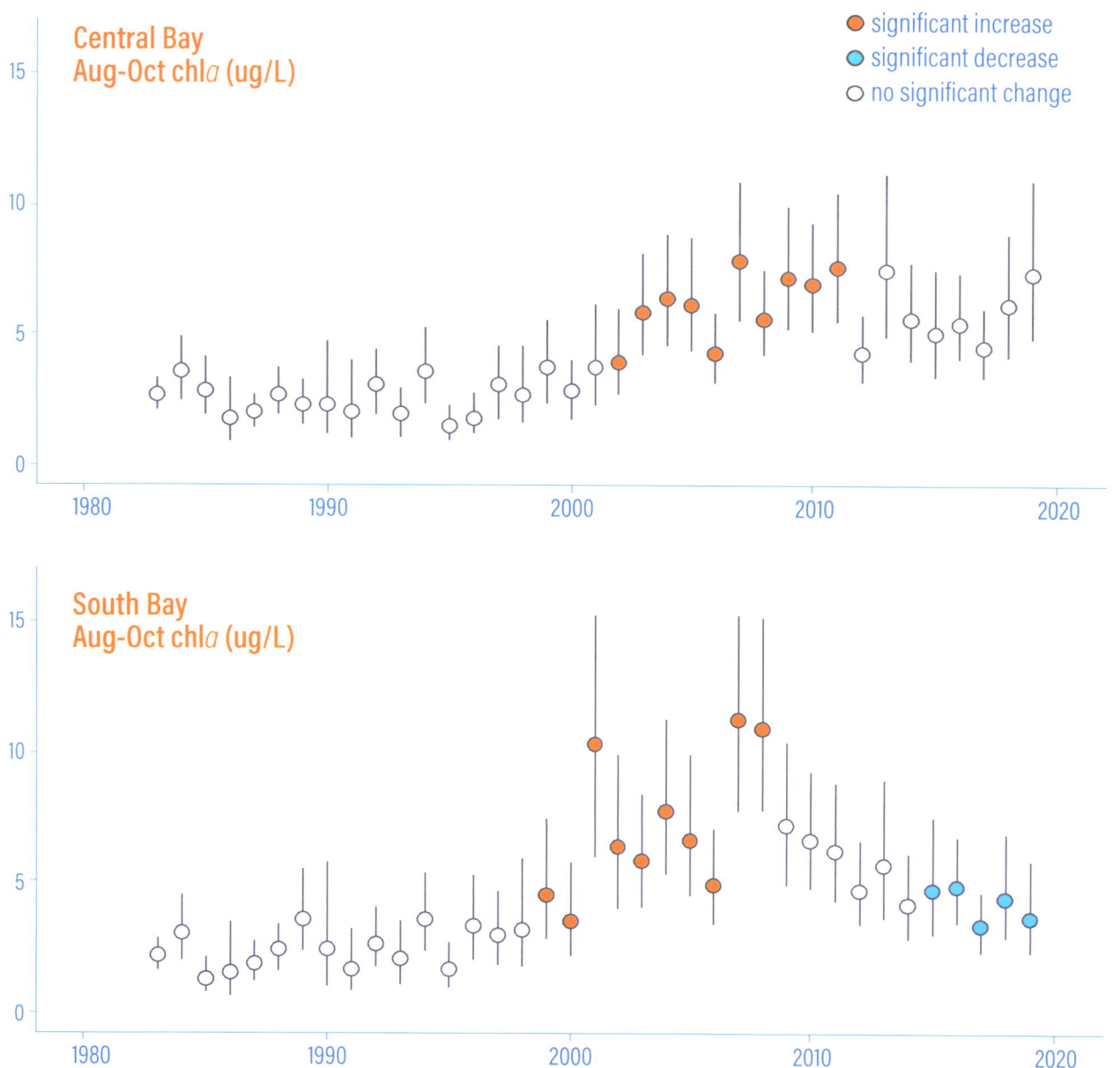

Estimated Mean Aug-Oct Chlorophyll Concentrations, 1983-2019

FOOTNOTE: **Estimated mean Aug-Oct chl*a* concentrations, 1983-2019** (vertical lines: 95% confidence intervals), in South Bay (s30, midway between the San Mateo and Dumbarton Bridges) and Central Bay (s21, near Bay Bridge). Symbol color represents long-term trend in Aug-Oct chl*a*, based on an 11-yr rolling window (right justified). Visit this webtool to explore long-term trends in chl*a*, dissolved oxygen, and gross primary productivity at South Bay and Central Bay stations. For additional information on the approach see Beck et al. (2021) and SFEI (2022). Data: USGS (Schraga and Cloern 2017).

Estimated Mean Aug-Oct Dissolved Oxygen Concentrations, 1993-2019

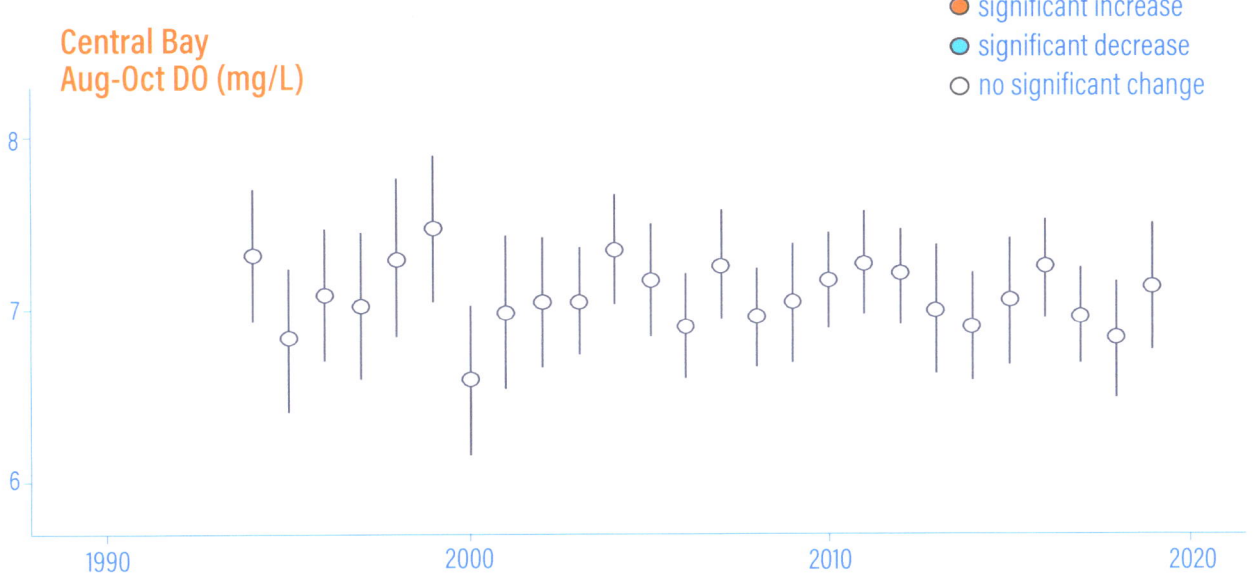

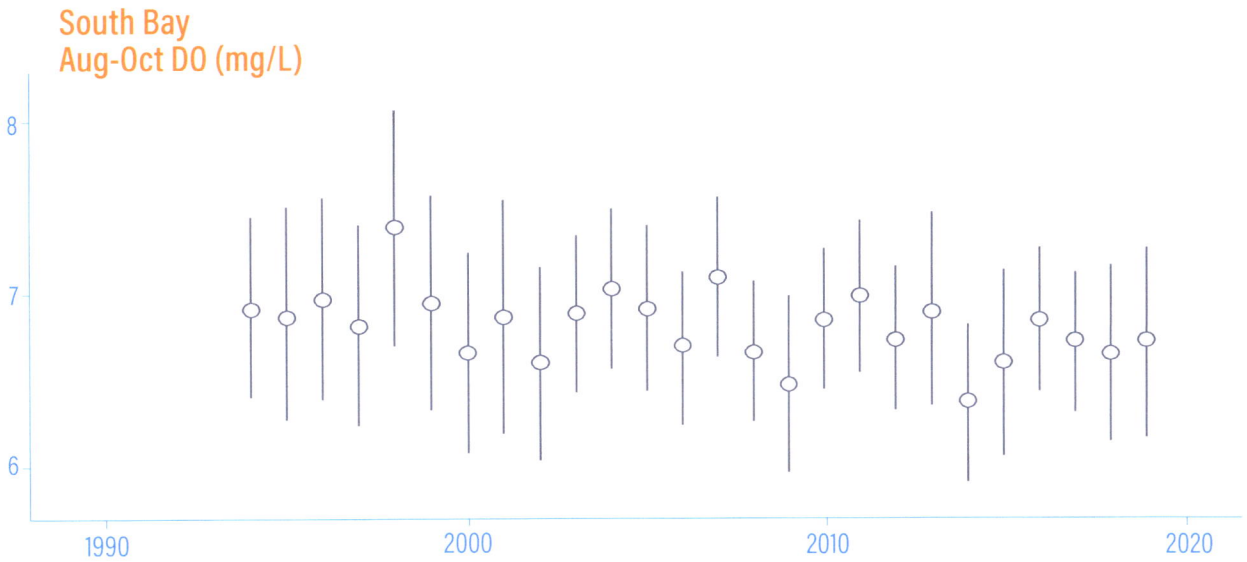

FOOTNOTE: **Estimated mean Aug-Oct depth-averaged dissolved oxygen concentrations, 1993-2019** (vertical lines: 95% confidence intervals), in South Bay (s30) and Central Bay (s21). Symbol color represents long-term trend in Aug-Oct DO, based on an 11-yr rolling window (right-justified; same legend as for chl*a*). For additional information on the approach see Beck et al. (2021) and SFEI (2022). Data: USGS (Schraga and Cloern 2017)

MERCURY

Origins of the Problem

The Bay's mercury problem has deep roots in California history, originating with the Gold Rush of 1849 that led to a rapid increase in population in the Bay Area and beyond, and statehood for California in 1850. Mercury was used in the process of separating gold from gold-bearing ore or sediment (placer) deposits. Lode and hydraulic gold mining in the Sierra Nevada combined to release a total of 3.6 to 6.0 million kg of mercury to the environment. The mercury used in gold mining primarily came from the rich deposits in the Coast Range of northern California. Between 1846 and 1981 approximately 104 million kg of mercury was extracted in California —88% of the mercury extracted in the entire US—and much of this production was from northern California Coast Range counties. An estimated 34 million kg of mercury was lost to the environment during ore processing ("furnace losses"). The most noteworthy mercury mine was New Almaden in the Guadalupe River watershed, the most productive mining district in the entire US. Contaminated tailings, soils, and drainage from historic gold and mercury mining districts have caused mercury contamination of downstream water bodies throughout the Bay watershed, and continue to supply contaminated sediment to the Bay today.

History of Monitoring

In 1969, as the scope of worldwide environmental contamination due to mercury was first being discovered, two striped bass from the Delta were found to have mercury in their muscle tissue at levels of concern (0.70 parts per million, or ppm). In 1970, as a result of this finding, an Interagency Committee was created to evaluate mercury contamination in California. The Committee initiated further studies of mercury in sport fish, commercial fish, game birds, water, and sediment. In samples collected between April and July 1970, 55 of 102 striped bass collected in the Bay-Delta were higher than 0.5 ppm. In late 1970, based on these studies, the first fish consumption advisory was issued for the Bay and Delta advising pregnant women and children not to consume striped bass. Limited additional monitoring of mercury in Bay fish occurred until 1994, when the San Francisco Bay Water Board conducted a comprehensive Bay-wide pilot study for mercury and other contaminants. In 1997, the RMP followed up on the pilot study and began a long-term monitoring effort that has continued to the present.

Mercury concentrations in Bay fish are well above the fish tissue objective and not showing signs of long-term decline

Advisories and Regulation

In 1993 the advisory for striped bass was revised by the California Environmental Protection Agency's Office of Environmental Health Hazard Assessment (OEHHA) to include size-specific consumption advice for adults, children 6-15 years, and pregnant women and children under age 6. In 1994, following the Water Board's pilot monitoring study, OEHHA issued an interim consumption advisory for multiple Bay species and multiple contaminants. In 2008, after several years of development, a Bay-wide total maximum daily load control plan (TMDL) and site-specific water quality objectives (based on mercury concentrations in fish) received final approval. Another mercury TMDL focused on the Guadalupe River watershed was approved in 2010. In 2011, based largely on the extensive dataset on mercury and other contaminants in Bay fish generated by the RMP, OEHHA issued a final Bay-wide advisory for mercury and other contaminants.

Current Status and Long-term Outlook

Mercury concentrations in Bay fish are well above the fish tissue objective and not showing signs of long-term decline. Striped bass remains the most important indicator species for mercury in the Bay, due to its popularity for consumption and the high concentrations that it accumulates. Striped bass from the Bay have the highest average mercury concentration measured for this species in US estuaries. The relatively extensive historical dataset Bay striped bass allows for the evaluation of trends over 44 years, from 1971 to 2019. In 2019, the average mercury concentration was not significantly different from the average in 1971. Furthermore, the overall long-term trend line does not indicate a change over the 44-year period. The primary source of mercury in Bay fish is sediment that was contaminated by historic mining activity and is now trapped in the Bay. Mercury levels in Bay fish can be expected to remain above thresholds of concern for decades to come. §

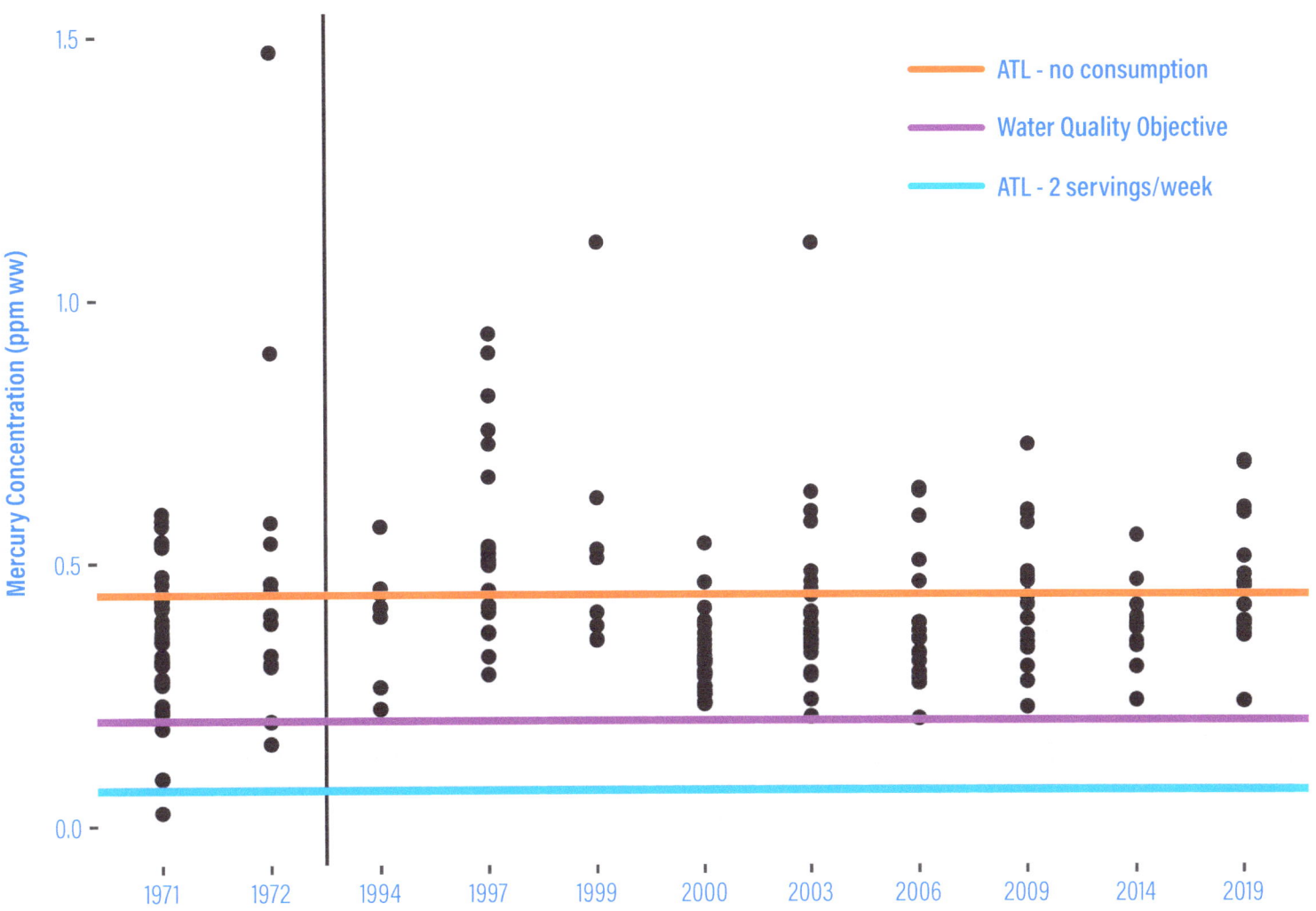

Mercury in Striped Bass, 1971-2019

FOOTNOTE: Bars indicate average concentrations. Points represent individual fish, with the exception of six composite samples (3 fish each) analyzed in 2014. All plotted points are 60 cm length-adjusted. The 2014 data do not include fish collected in Artesian Slough, and the 2019 data do not include fish collected in South Bay (Coyote Creek); these areas reflect unique mercury sources and were collected only in those years. Data were obtained from CDFW historical records (1971-1972), the Bay Protection and Toxic Cleanup Program (1994), a CalFed-funded collaborative study (1999 and 2000), and the Regional Monitoring Program (1997, 2000, 2003, 2006, 2009, 2014, and 2019). The colored lines indicating advisory tissue level (ATL) thresholds show the lower end of ATL ranges for the sensitive population.

PCBs

Origins of the Problem

Polychlorinated biphenyls (PCBs) are extremely persistent synthetic chemicals that were heavily used from the 1930s to the 1970s in hundreds of industrial and commercial applications including electrical, heat transfer, and hydraulic equipment; pigment, dye, and carbonless copy paper; and in plasticizer applications in paint, plastic, and rubber. PCBs were used as plasticizers in products where high durability was a requirement such as in industrial grade paint coatings around high-voltage wiring, and in caulking compounds most commonly used in commercial, industrial, and institutional concrete and masonry structures, including buildings, dams, airport runways, bridges, foot paths, parking structures, wastewater treatment plants, storm drains, and roads. Awareness of their presence in the environment and their toxicity to humans and wildlife grew in the 1960s and 1970s, leading to federal legislation in 1979 to ban their production and sale. After decades of PCB use, San Francisco Bay has been left with a legacy of contamination spread widely across the land surface of the watershed, mixed deep into the sediment of the Bay, and moving through the Bay food web.

History of Monitoring

The first measurements of PCBs in the Bay were made in shiner surfperch collected in 1965. The mean concentration measured in three composite samples was 830 ppb. The first sustained annual monitoring of PCBs in the Bay was initiated by the State Mussel Watch Program in 1980, so trends during the critical period of the 1970s when use reductions began are unclear. Mussel monitoring continued through the 2000s, and provided some indications of declines. However, after the San Francisco Bay Water Board conducted a comprehensive Bay-wide pilot study for PCBs, mercury, and other contaminants in 1994, fish became the key indicator of PCB impairment of Bay water quality. In 1997, the RMP followed up on the pilot study and began a long-term monitoring effort that has continued to the present.

History of Advisories and Regulation

The most important management actions ever taken to reduce PCB contamination in the Bay were the phaseout during the 1970s and the 1979 federal ban on production and sales. The 1979 legislation, however, still allowed continued use of PCBs in existing equipment, and because of this a substantial stock of PCBs remains in use today. From the 1980s to the present, additional management of PCBs in the Bay has been driven by regulations pertaining to the cleanup of highly contaminated sites in the watershed and in the Bay. Some PCB-contaminated sites have been remediated under the Comprehensive Environmental Response, Compensation, and Liability Act (CERCLA), or "Superfund." In 1994, in response to the Water Board pilot fish monitoring study, the California Office of Environmental Health Hazard Assessment (OEHHA) issued an interim fish consumption advisory for all of San Francisco Bay. This led to the inclusion of the Bay on the 1998 California 303(d) List of impaired water bodies, which in turn led to the development of a Bay-wide TMDL control plan that was ultimately approved in 2010. The TMDL established PCB concentrations in sport fish as the key indicator of impairment. In 2011, based largely on the extensive fish contamination dataset generated by the RMP, OEHHA issued a final Bay-wide consumption advisory for PCBs and other contaminants.

Current Status and Long-term Outlook

More than forty years after the ban, PCBs are still far higher than the fish tissue target of 10 ppb established by the TMDL and have not shown clear signs of decline over the last 20 years. Shiner surfperch is the main indicator species, and had a Bay-wide average concentration of 220 ppb in the most recent sampling in 2019. This is lower than the concentration observed back in 1965, but surprisingly not that much lower. PCB concentrations in shiner surfperch across five long-term monitoring locations were generally higher in 2019 than in the prior round of sampling, but there are some possible signs of long-term decline. Overall, the rate of PCB decline in the Bay is slow at best, and continued monitoring is needed for a more definitive assessment.

Detailed studies of PCBs have been conducted at selected locations and provide some indication that concentrations in Bay fish on a local scale could decline in response to reductions in loads from nearby watersheds. More rigorous modeling and monitoring are needed to better evaluate this forecast. §

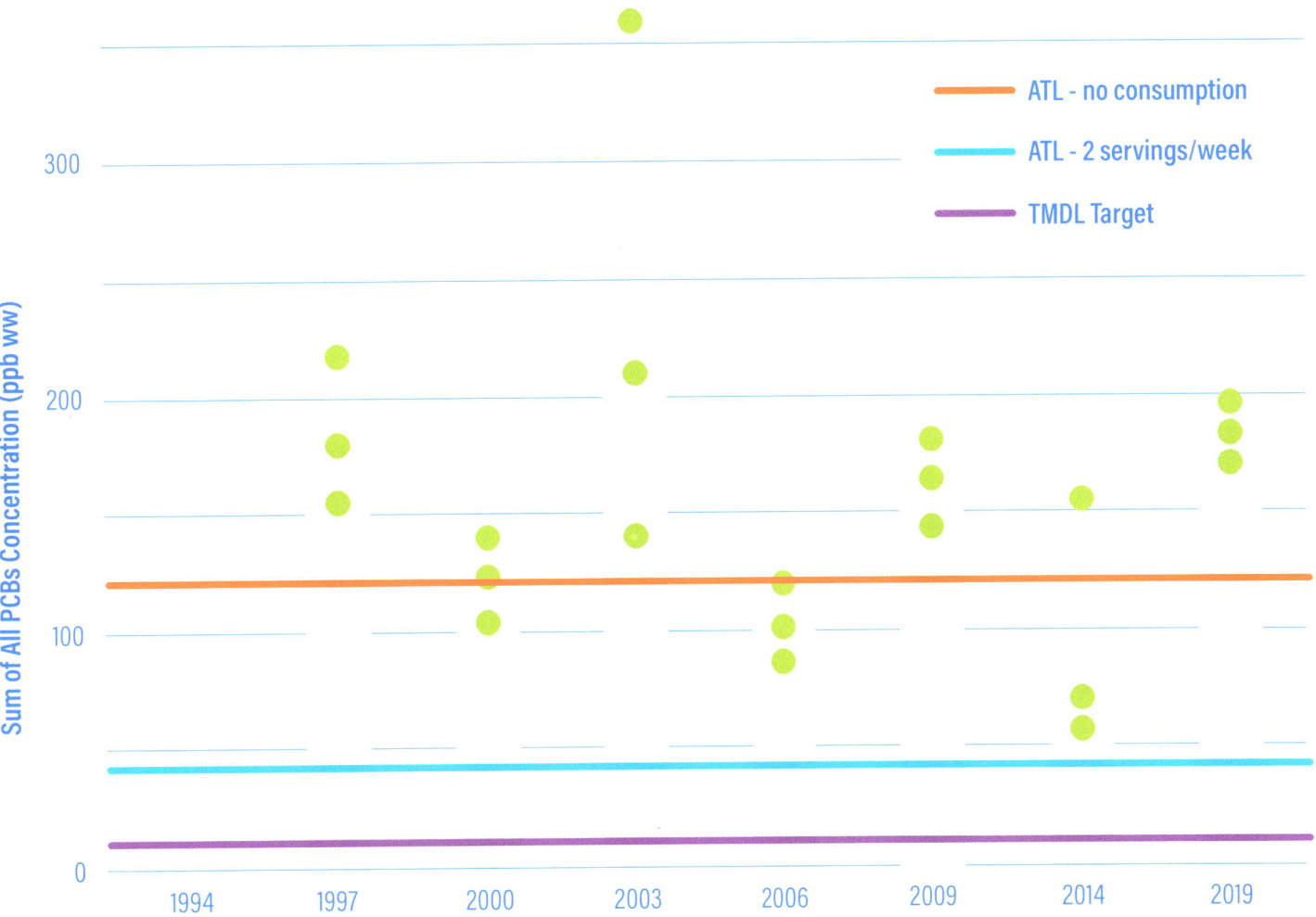

FOOTNOTE: Bars indicate average concentrations. Points represent composite samples with 20 fish in each composite. Data shown are the sum of PCBs for all congeners analyzed; the number analyzed varied from 47 in 1994 to 52 in 2014, and then increased to 209 in 2019. The colored lines indicating advisory tissue level (ATL) thresholds show the lower end of the ATL ranges.

WATER QUALITY PARAMETER SUMMARIES

DIOXINS

Origins of the Problem

Polychlorinated dibenzodioxins and dibenzofurans (commonly referred to as "dioxins") are highly toxic, persistent, and have a strong tendency to accumulate in the food web. Dioxins are mostly formed as byproducts of combustion of various materials and of manufacturing processes using chlorine such as pulp bleaching and production of polyvinyl chloride. In the past, "point source" emissions from facilities such as incinerators and smelters were thought to be the largest sources. As national regulation of dioxins tightened, most of these large point sources have been controlled. Today the main sources and pathways to the Bay are generally more dispersed, and include cars and trucks, residential wood combustion, wildfires, historically deposited residues in the environment, municipal wastewater treatment plants, and industrial discharge.

History of Monitoring

Accumulation in fish and the potential for human exposure through fish consumption is the primary concern driving dioxin monitoring and regulation. In 1994 the San Francisco Bay Water Board included dioxins in their comprehensive Bay-wide fish monitoring pilot study. Dioxins were detected at levels of concern, so monitoring continued in 1997 and beyond when the RMP followed up on the pilot study and began the long-term fish monitoring that has continued to the present. In the 2000s and 2010s, the RMP also evaluated dioxins in other matrices (bird eggs, water, surface sediment, sediment cores, wastewater, and stormwater) to address questions regarding long-term trends, spatial patterns, and pathways for input to the Bay. The bird egg monitoring has indicated significant declines and is continuing in order to provide a more definitive assessment.

Advisories and Regulation

In the wake of the findings of the Water Board's 1994 fish monitoring pilot study, OEHHA immediately issued a new interim fish consumption advisory for the Bay. The advice was issued due to concern over human exposure to residues of methylmercury, PCBs, dioxins, and organochlorine pesticides in Bay-caught fish. In 1998 USEPA included the Bay on the 303(d) List for dioxins based on their inclusion in the interim consumption advisory and USEPA's own assessment of the available data indicating potential health risk to consumers. The Bay continues to be listed for dioxins today. Although dioxins are at concentrations of potential concern in the Bay, neither a Water Board regulatory target nor OEHHA advisory tissue levels have been established. As part of the PCB TMDL, because some PCBs have the same mechanism of toxicity as dioxins, the Water Board calculated a fish tissue screening level for dioxins of 0.14 pptr (parts per trillion) for the assessment of risk to human health. Dioxin concentrations in Bay fish have consistently exceeded this screening level.

Current Status and Long-term Outlook

Dioxin concentrations in Bay fish remain above the Water Board screening level, and are still particularly high in Oakland Harbor. However, there are signs of possible decline in both of the key indicator species: shiner surfperch and white croaker.

In shiner surfperch, concentrations appear to be progressively decreasing across all of the monitoring stations sampled around the Bay except Oakland Harbor, although the decline is not statistically significant at any of the monitoring stations. In white croaker (data not shown), the concentrations in 2019 were sharply lower than the last year of comparable data in 2009, and only slightly above the screening level. Continued monitoring of shiner surfperch and white croaker is needed to establish whether these possible trends reach a point of statistical significance and are signs of actual long-term declines.

The datasets for fish and bird eggs suggest dioxin concentrations have declined in the Bay. Several recent management efforts not specifically aimed at dioxins should contribute to decreased loading to the Bay, including Bay Area Air Quality Management District bans on wood-burning devices in new homes and rebates for conversion to natural gas fireplaces, efforts by local municipalities to reduce PCB runoff via green stormwater infrastructure, and incentives to shift from gas-powered to electric vehicles. On the other hand, the extensive wildfires that have occurred more frequently in recent years are expected to generate dioxins and, if they occur in the local Bay watershed, could counterbalance declining inputs from other sources. §

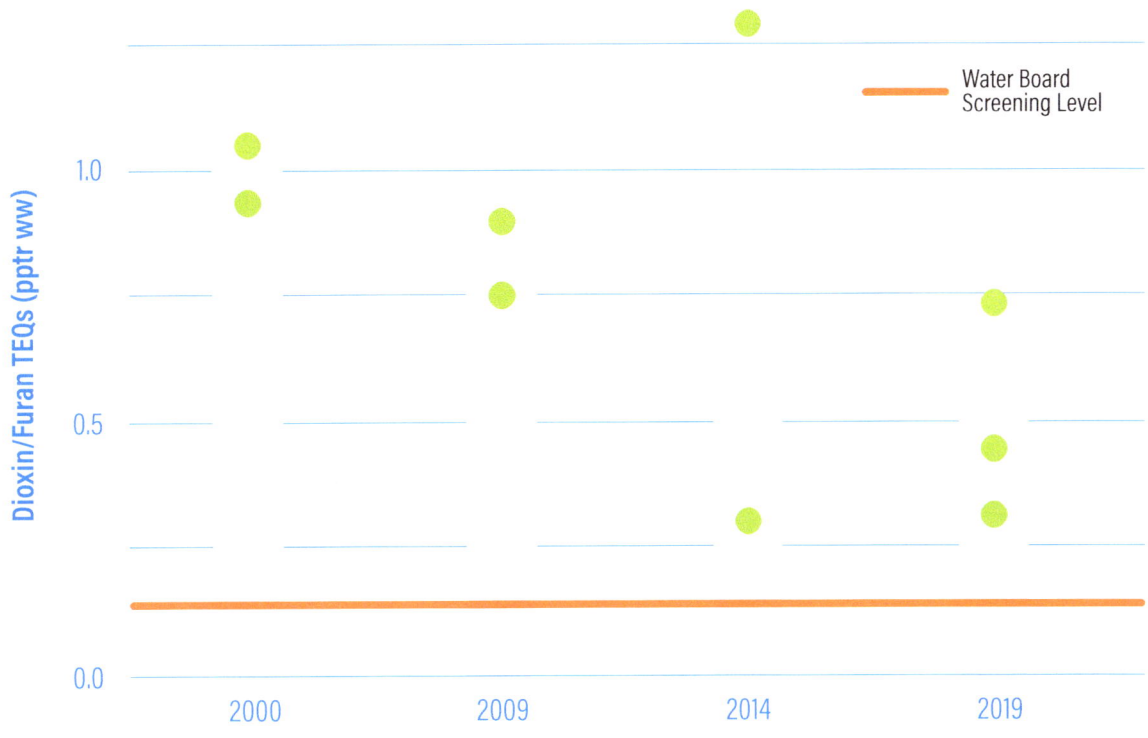

Dioxins in Shiner Surfperch, San Francisco Waterfront, 2000-2019

FOOTNOTE: Bars indicate average concentrations. Points represent composite samples with 20 fish in each composite. TEQ = toxic equivalent.

SELENIUM

Origins of the Problem

Selenium in trace amounts is essential for cellular function in many organisms, including all animals. Fossil fuels like crude oil are derived from living organisms and therefore contain selenium. Certain rock formations, such as marine shales in the western San Joaquin Valley, also have naturally elevated selenium concentrations due to the presence of fossilized marine biota.

Selenium enters the Bay from a broad range of pathways, including rivers, stormwater, municipal and industrial wastewater, and atmospheric deposition. The two largest inputs are from the San Joaquin River and oil refinery discharge, both of which are pathways into northern San Francisco Bay (San Pablo Bay and Suisun Bay). Selenium export from the San Joaquin River watershed is high due to the geologic sources and agricultural practices that exacerbate its release from soils. The amount of selenium delivered to the Bay via the San Joaquin River is highly variable and dependent on the amount and intensity of rainfall within a year. The five oil refineries in the North Bay discharge a more consistent load.

While selenium is an essential micronutrient, it is also toxic to aquatic life at levels minimally higher than what is essential. Selenium toxicity mainly occurs via consumption of contaminated prey rather than uptake from the water column. In San Francisco Bay, the invasive clam *Potamocorbula amurensis* accumulates selenium at a higher rate than native clams. The spread of this invasive clam and its efficient uptake of selenium is largely responsible for the accelerated bioaccumulation of selenium in the food web, particularly for bird and fish species that feed on *Potamocorbula,* such as white sturgeon and surf scoters.

History of Monitoring

Analysis of selenium in surf scoters and greater scaups collected from South San Francisco Bay in 1982 revealed that selenium concentrations were similar to those in ducks sampled in the San Joaquin Valley where reproduction was severely impaired. This finding and studies that followed prompted the establishment of a USGS-led long-term monitoring program for *Potamocorbula* in the North Bay that ran from 1995 to 2017. Monitoring for dissolved, particulate, and total selenium in the water column has been less consistent and widespread, with RMP data extending back to 1993; USGS sampling from 2007 to 2017; and limited additional sampling in 1999-2000, 2010, and 2012. White sturgeon monitoring has been conducted by the RMP since 1994, with samples collected every three years from 1994-2009 and every five years from 2014 onwards. The RMP also piloted a muscle plug monitoring program from 2015-2017. Additional data were collected through the State Water Board's Selenium Verification Study (1986-1990), by the USGS during sturgeon derbies (1999-2001), and by UC Davis, California Department of Fish and Wildlife, and the Bureau of Reclamation from 2002-2005.

History of Advisories and Regulation

North San Francisco Bay was identified as being impaired for selenium in 1998 based on elevated concentrations in ducks and associated human health consumption advisories (first issued in 1987) and concerns for impacts on the ducks themselves and on white sturgeon. In 2000, the USEPA issued water quality standards for priority toxic pollutants for enclosed bays and estuaries in California, which included selenium. These standards—referred to as the California Toxics Rule—included water quality criteria for selenium that were not specific to the San Francisco Bay and Delta. During consultation with the US Fish and Wildlife Service and National Marine Fisheries Service, a revision to the selenium water quality criteria was recommended, as well as the development of criteria that would be more protective of aquatic life. In July 2016 the USEPA proposed aquatic life and aquatic-dependent wildlife criteria for selenium in the Bay and Delta. A selenium TMDL for North San Francisco Bay was developed by the San Francisco Bay Water Board and approved in 2016, establishing numeric targets for selenium in fish tissue and total dissolved selenium in the water column based on bioaccumulation models specific to the Bay.

Current Status and Long-term Outlook

Selenium loading from the Delta decreased one-half to two-thirds between the mid-1990s and mid-2000s in large part due to changes in agricultural practices in the San Joaquin Valley. Selenium concentrations on the San Joaquin River continued to decline through 2016, likely due to the control efforts of the Grassland Bypass Project. In the late 1990s, loading from the Bay Area refineries decreased 75% due to selenium removal measures added to the wastewater treatment process. Concentrations in white sturgeon were highest in 1990, and were generally lower for the rest of the 1990s and 2000s but still hovered near the TMDL numeric target. Higher concentrations in sturgeon tissue were recorded again in 2014-2016 in the last three years of a five-year drought, followed by lower concentrations in 2017 after a wet winter. For clams, there has been no significant trend in selenium concentrations in the long-term time series, suggesting that loading decreases to date have not had a marked effect on Bay biota. One of the biggest factors determining selenium concentration in sturgeon and clams is the amount of Delta outflow entering North San Francisco Bay, with lower concentrations in wet years. While loads have decreased and are not expected to increase in the years to come, other factors such as drought and alterations in the hydrology of the Delta could potentially lead to changes in selenium concentrations in the North Bay food web. §

Selenium in White Sturgeon, 1987-2019

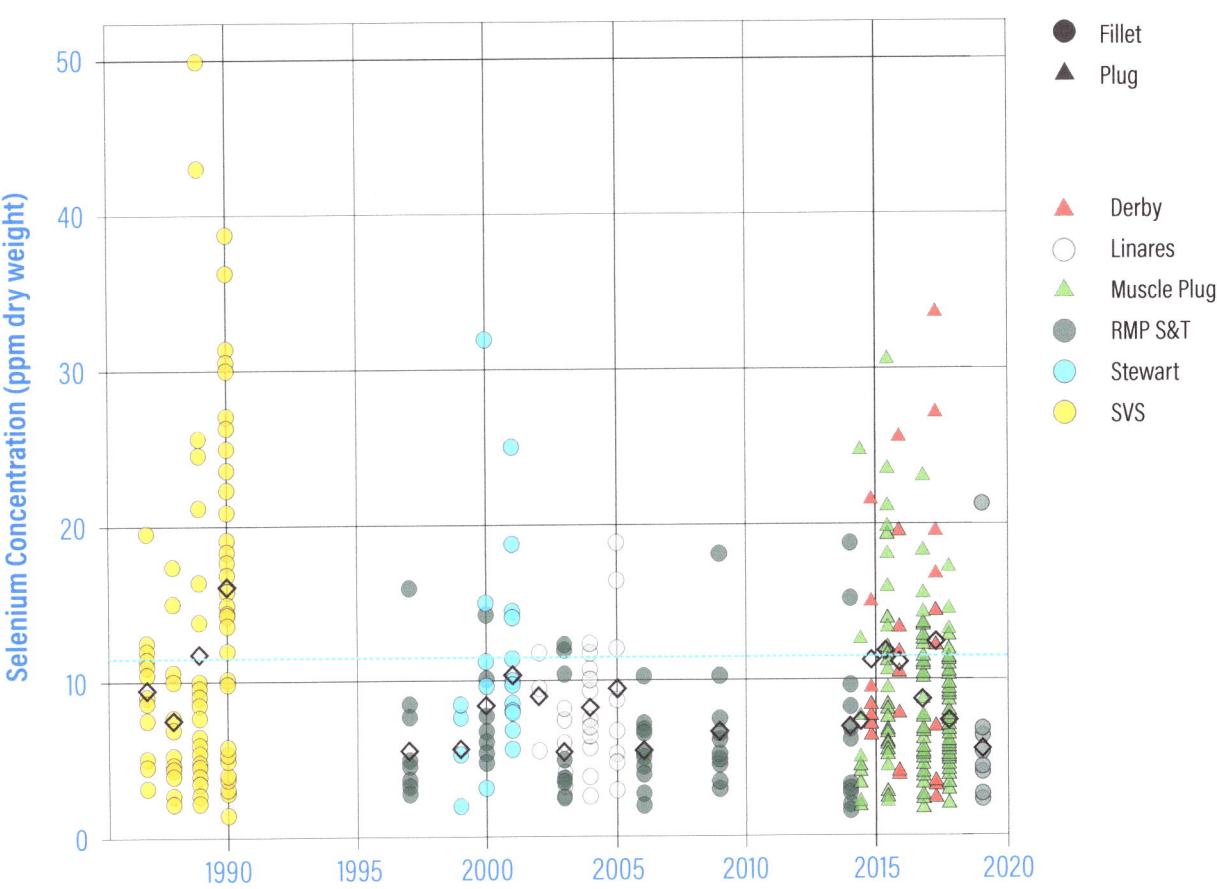

FOOTNOTE: Points represent samples of individual white sturgeon. Mean concentrations for each study, or each year of multi-year studies, are shown in black diamonds. Horizontal blue line indicates the North Bay TMDL target for selenium in sturgeon muscle tissue (11.3 µg/g dw). Data from the RMP and other sources as follows: Derby – Sun et al. (2019); Linares – Casenave et al. (2015); Muscle Plug – Sun et al. (2019b); RMP S&T (1997- 2014); Stewart – Stewart et al. 2004; SVS (Selenium Verification Study) – Urquhart et al. 1991.

COPPER

Origins of the Problem

After the great progress in managing oxygen-depleting organic waste and bacteria in the initial decade of Clean Water Act implementation, attention in the 1980s began to turn to toxic contaminants, with an initial focus on toxic metals. From the Gold Rush to the 1950s, toxic contaminant inputs to the Bay generally increased along with the growth of the population and the flow of untreated sewage. The construction of sewage treatment plants that began in the 1950s and culminated in nearly all sewage receiving secondary treatment by 1980 also greatly reduced toxic metal inputs, but concerns still remained. Copper concentrations in South Bay clams measured in the 1970s were among the highest ever observed in estuarine benthos. Water monitoring that began in earnest in the late 1980s found concentrations that were elevated relative to criteria available at that time. Copper is of much less concern today due to improvements in understanding based on robust monitoring and other studies.

History of Monitoring

There were limited data on toxic metal concentrations in Bay waters prior to the 1990s, and even those few measurements were of questionable accuracy. In 1989 the Water Board contracted with Dr. Russ Flegal at UC Santa Cruz to monitor metals in the Bay using ultra-clean methods, generating for the first time a reliable Bay-wide dataset to compare to water quality objectives. This pilot work was the precursor of the RMP and demonstrated that cost-effective regional monitoring to address management questions was possible. When the RMP began in 1993, it continued to contract with the Flegal lab to perform Bay-wide metals monitoring. In addition to the RMP monitoring, many other studies substantially improved understanding of the cycling and bioavailability of metals in the Bay in the 1990s. As a result, by the early 2000s the Bay had become one of the most extensively studied estuaries, if not the most extensively studied estuary, in the world for trace metals. After 2011, the list of metals analyzed in water by the RMP was scaled back to only copper, methylmercury, and selenium because data for the other metals were not addressing high priority management questions. After 2019, copper became the only metal with continued monitoring.

History of Advisories and Regulation

The first 303(d) listings of the Bay for toxic pollutants came in the early 1990s when all Bay segments were listed as impaired by metals in Bay waters. These metals listings were subsequently refined in 1996 to just copper, mercury, nickel, and selenium.

An impairment assessment for copper by stakeholders and the Water Board, based on an extensive dataset provided by the RMP and other studies showing that most of the copper in the Bay is bound up in a harmless form, concluded that the existing water quality objectives were inappropriately low. These findings led to new Bay-specific water quality objectives for copper (less stringent but still considered fully protective of aquatic life), pollution prevention and monitoring activities to make sure concentrations remain below the objectives, and the 2002 removal of copper from the 303(d) List of pollutants of concern in the Bay.

To maintain water quality in the Bay, municipalities are required to implement actions to control discharges to storm drains from architectural (e.g., roofs) and industrial (e.g., metal plating) uses of copper, as well as copper used as an algaecide in pools, spas, and fountains. They are also required to address vehicle brake pads, the largest source of copper to the Bay, which they have done through participation in the Brake Pad Partnership, a public-private collaboration whose work led to the passage of legislation (SB 346) requiring that the amount of copper in brake pads sold in California be reduced to no more than 0.5% by 2025.

In order to determine that concentrations have not increased, monitoring data collected by the RMP are compared to site-specific trigger levels. If the trigger concentration is exceeded in any Bay segment, the Water Board will investigate causes of the exceedance and consider potential control options.

Current Status and Long-term Outlook

Different trigger levels have been established for each Bay segment (depicted on the graphs). Rolling averages covering three sampling rounds are compared to the triggers. Across all of the Bay segments, these averages have always been below the triggers. In only a few instances did the upper 95% confidence intervals of these rolling averages exceed the trigger value (2009, 2011, 2013, and 2015 in Lower South Bay). In the most recent rounds of sampling, the averages have been the lowest or among the lowest measured across the entire period of record, and even the upper end of the 95% confidence intervals have been well below the triggers. Additional rounds of sampling are needed to definitively establish whether these recent results are indicative of a long-term decline. With the reduction of the largest source of copper to the Bay (brake pads) is well underway toward the 2025 virtual copper phase-out, the expectation is that copper concentrations will indeed show such a decline. §

Copper in Bay Water, 2002-2021

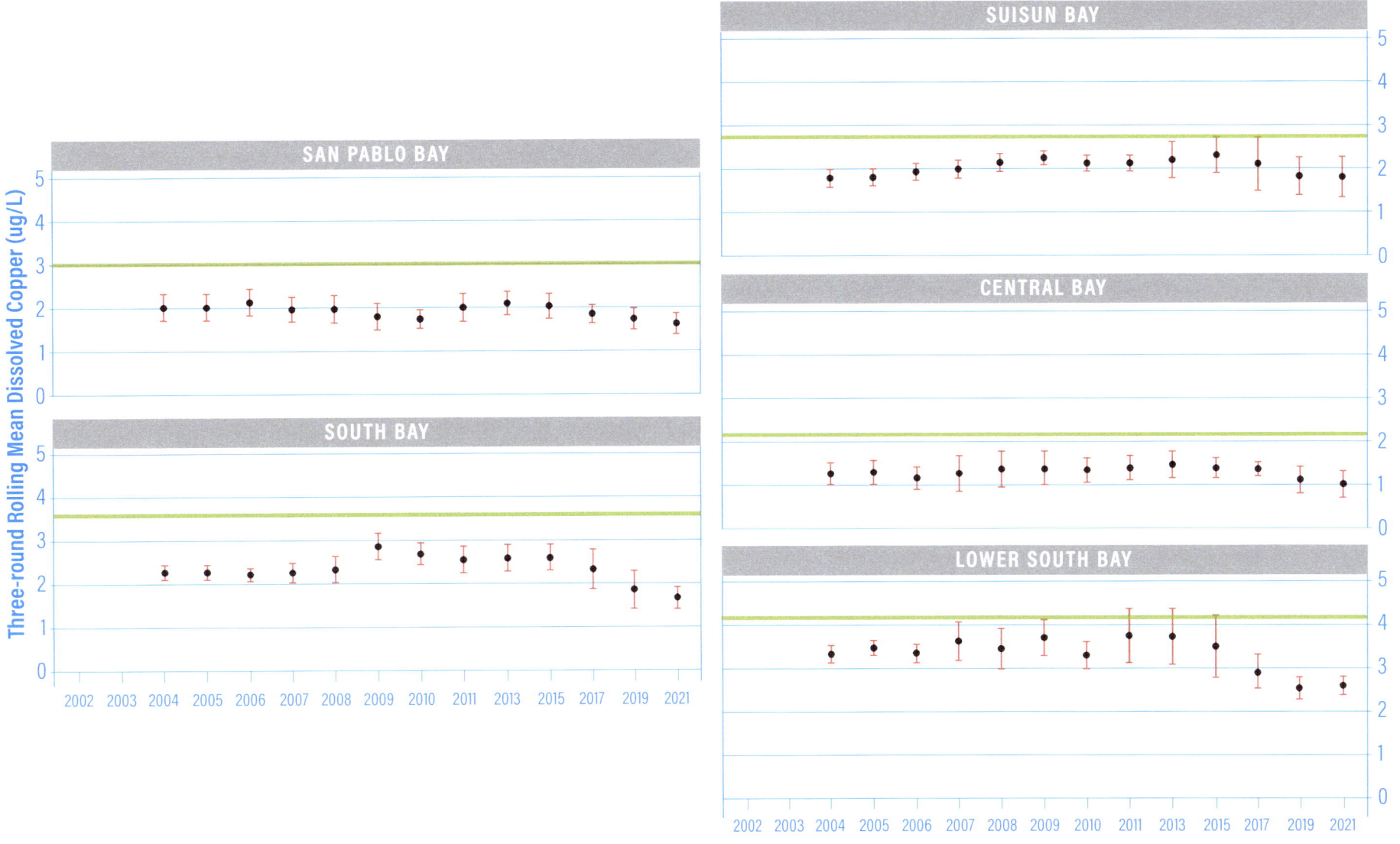

FOOTNOTE: Trend plots show annual random-station three-round rolling means for dissolved copper with error bars indicating the 95% confidence intervals of the means. Lines show the trigger values for each segment.

WATER QUALITY PARAMETER SUMMARIES

PBDEs

Origins of the Problem

In the 1970s, California established flammability standards for a variety of consumer goods such as upholstered furniture and products for infants and young children. Our state standards became de facto flammability standards across the nation.

To meet these standards, manufacturers began to add large quantities of chemical flame retardants including polybrominated diphenyl ethers (PBDEs) to their products. Three commercial formulations of PBDEs were widely used for decades. An average couch manufactured to meet the state standard contained about two pounds of PBDEs.

History of Monitoring

Extensive use in consumer goods led to PBDE exposure in San Francisco residents and wildlife. A 2002 California Department of Toxic Substances Control study reported high levels of PBDEs in blubber from Bay harbor seals, with data from archived samples suggesting that concentrations had doubled every 1.8 years throughout the 1990s. A study on Forster's tern eggs collected in 2002 revealed the highest levels of PBDE contamination in biota reported at the time, 63,000 ppb on a lipid weight basis. This concentration remains one of the highest ever recorded in any organism. The DTSC research team also found high levels of PBDEs in tissue samples from Bay Area women. At the same time, significant concerns about the impacts of PBDEs to human and ecological health began to emerge.

The RMP began monitoring PBDEs in 2002 in a variety of Bay matrices, with particular focus on sediment and biota including bivalves, bird eggs, and sport fish. The RMP also conducted a special study to examine these bioaccumulative contaminants in harbor seal blubber collected in 2014, and compared results to earlier findings for this species.

History of Advisories and Regulation

In response to toxicity concerns and rapidly increasing concentrations in humans and wildlife, the California Legislature banned two types of PBDE mixtures in 2006, leading to a nationwide phase-out. US chemical manufacturers announced a halt in production of the last PBDE mixture in 2013. Meanwhile, in 2011, the California Environmental Protection Agency's Office of Environmental Health Hazard Assessment (OEHHA) established advisory tissue levels (ATLs) for PBDEs in sport fish. Previous rounds of RMP monitoring showed that PBDE concentrations in Bay fish were well below the two serving per week ATL of 100 ppb wet weight.

Decades after creation of the initial flammability standard for furniture and children's products, California state scientists determined that the standard did not actually provide significant protections in the event of a fire. In 2013 the state developed a new standard for these products (California Technical Bulletin 117-2013) that was protective and did not require use of chemical flame retardants. The federal government later adopted California's revised standard. More recently, in 2018 California passed AB 2998, a ban on all flame retardants in upholstered furniture, mattresses, and children's products, effective in 2020.

Current Status and Long-term Outlook

The ban and phase-out of PBDEs succeeded in driving a rapid and substantial drop in concentrations in the Bay. Two decades of RMP monitoring indicates significant PBDE declines in multiple matrices, including Bay cormorant eggs and many others. In 2017, the RMP documented sufficient recovery to move PBDEs from moderate to low concern in the tiered, risk-based framework for contaminants of emerging concern in the Bay.

> *The ban and phase-out of PBDEs succeeded in driving a rapid and substantial drop in concentrations in the Bay*

At the same time, RMP special studies have documented the presence of alternative flame retardants, such as organophosphate esters (page 70, in Bay water, sediment, and biota. Flame retardants like these are still widely used in a number of product categories, such as electronics and building insulation, and may pose risks to aquatic life. §

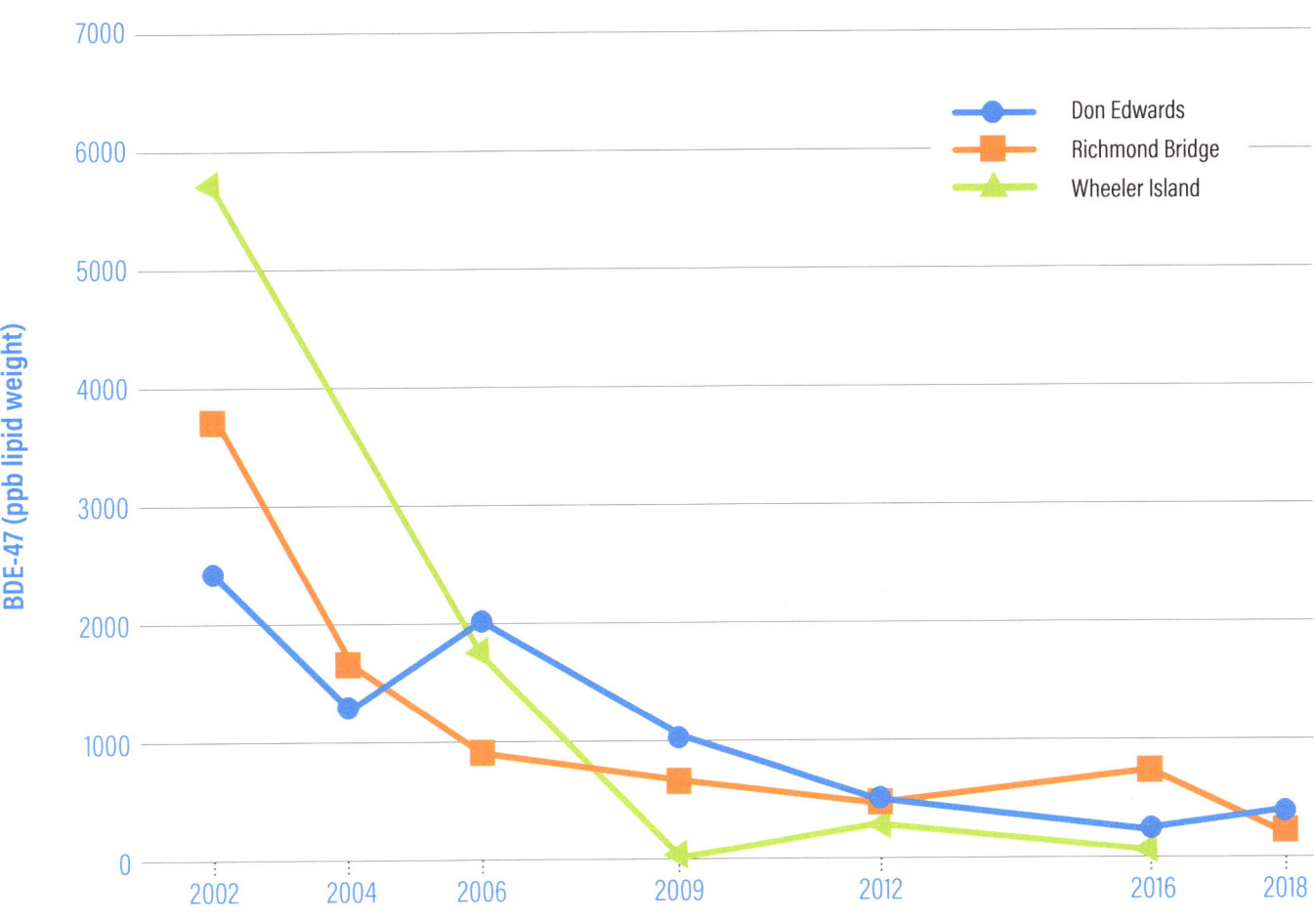

PBDEs in Cormorant Eggs

FOOTNOTE: Average BDE-47 concentrations (ppb lipid weight) in cormorant egg composites. BDE-47 is the most abundant BDE congener in the eggs and is presented as an index of PBDEs as a whole. Each point represents three composites, with 7 eggs in each composite. The Suisun Bay (Wheeler Island) colony could not be sampled in 2018. Data from the next round of sampling in 2022 are not yet available.

PFAS

Origins of the Problem

Per- and polyfluoroalkyl substances (PFAS), a class of thousands of synthetic, fluorine-rich compounds commonly referred to as "forever chemicals," are known for their thermal stability, non-reactivity, and surfactant properties. Originally developed in the 1940s, these unique compounds have widespread uses across consumer, commercial, and industrial products including food packaging materials, waterproof textiles, stain-resistant carpets and furniture, personal care products, processing aids to produce fluoropolymers like Teflon, and hydraulic aviation fluids. PFAS are also well known for use in aqueous film forming foam (AFFF), initially created in the 1960s for fire suppression for the US military. Widespread use across consumer, industry, and military applications has resulted in contamination in San Francisco Bay and across the globe.

History of Monitoring

In the early 2000s, PFAS began to gain attention as sensitive analytical methods became widely available. Over the past two decades, PFAS contamination has been documented worldwide. Since 2004, the RMP has tracked PFAS in the Bay via a series of studies on harbor seals, cormorants, fish, bivalves, sediment, ambient water, wastewater, and stormwater.

Across Bay biota, PFAS are ubiquitous, especially those most extensively used historically: perfluorooctane sulfonate (PFOS) and perfluorooctanoate (PFOA). In 2004 and 2006, concentrations of PFOS in Bay harbor seals and bird eggs were some of the highest detected globally. Continued triennial bird egg monitoring has indicated decreases in PFOS, though current levels may still pose risks. Sport fish monitoring has shown the presence of PFOS at concentrations exceeding thresholds for consumption advisories for human health that have been established by other states, especially in South Bay fish.

PFAS have been observed in stormwater and wastewater, significant pathways to the Bay. Studies using advanced techniques have also indicated a significant fraction of PFAS discharged to the Bay is from unknown compounds that are not quantified by standard methods. A recent regional study of influent, effluent, and biosolids on behalf of the Bay Area Clean Water Agencies (BACWA) detected various PFAS, including unknown compounds, across each matrix.

History of Advisories and Regulation

To date, much of the regulatory focus has been on long-chain perfluoroalkyl chemicals such as PFOS and PFOA. In the US, production of PFOS was phased out by 2002, and production of PFOA was phased out by 2015. These federal actions were part of a broader international effort to reduce human and environmental risks associated with these compounds. This year, the USEPA has announced that it will propose designating PFOS and PFOA as hazardous substances under the federal Superfund program, a move that is expected to spur cleanup of numerous contaminated sites.

Drinking water contamination is a major focus of federal and state regulators. USEPA issued an initial drinking water lifetime health advisory for PFOA and PFOS in 2016. In June 2022, with increasing study of the dangers of PFAS, USEPA greatly reduced its health advisory levels to 0.02 ppt and 0.004 ppt for PFOA and PFOS, respectively, and released additional health advisories for a newly developed replacement PFAS (Gen-X) and for perfluorobutanesulfonate (PFBS). Similarly, the California State Water Resources Control Board (SWRCB) introduced its first drinking water notification and response levels for PFOA and PFOS in 2018. More recently, the SWRCB reduced these levels, and issued notification and response levels for PFBS.

The SWRCB has launched efforts to test drinking water, wastewater, and other matrices at numerous sites across the state, focusing on locations near airports, military bases, landfills, and other potential sources of PFAS to the environment. In 2020, the San Francisco Bay Regional Water Quality Control Board developed interim final Environmental Screening Levels (ESLs) for PFOS and PFOA in aquatic habitats based on ecotoxicity and seafood ingestion, as part of establishing broader interim guidance for groundwater and soil contamination. At this time, California does not provide tissue thresholds or guidance concerning consumption of sport fish contaminated with PFAS.

California has banned the use of firefighting foams with PFAS (SB 1044, 2020), with bans on PFAS in foodware (AB 1200, 2021) and children's products (AB 652, 2021) set to take effect in 2023. The state has also issued regulations intended to restrict the use of PFAS in carpets and rugs through the DTSC Safer Consumer Products Program.

Current Status and Long-term Outlook

PFAS are extremely persistent and can be strongly bioaccumulative, leading to build up over time in the Bay. Observations over time are consistent with shifts in manufacturing away from PFOS and PFOA and towards newer PFAS alternatives, which are also likely to pose risks to wildlife. RMP monitoring efforts are expanding to better characterize their occurrence and trends across biotic and abiotic matrices. A recent study quantified the presence of multiple PFAS in Bay water, including elevated levels in the South and Lower South Bay. With concentrations in fish exceeding protective consumption thresholds established by other states, a virtual forum in early 2022 marked the beginning of an effort to work with local community groups and stakeholders to protect fishing communities.

The widespread presence of PFAS across California has increased the urgency to address these compounds at the state and federal levels. State regulators and lawmakers continue to explore and implement solutions to protect the public. USEPA's PFAS Strategic Roadmap is a strong indication of further standards and regulations to come at the federal level. §

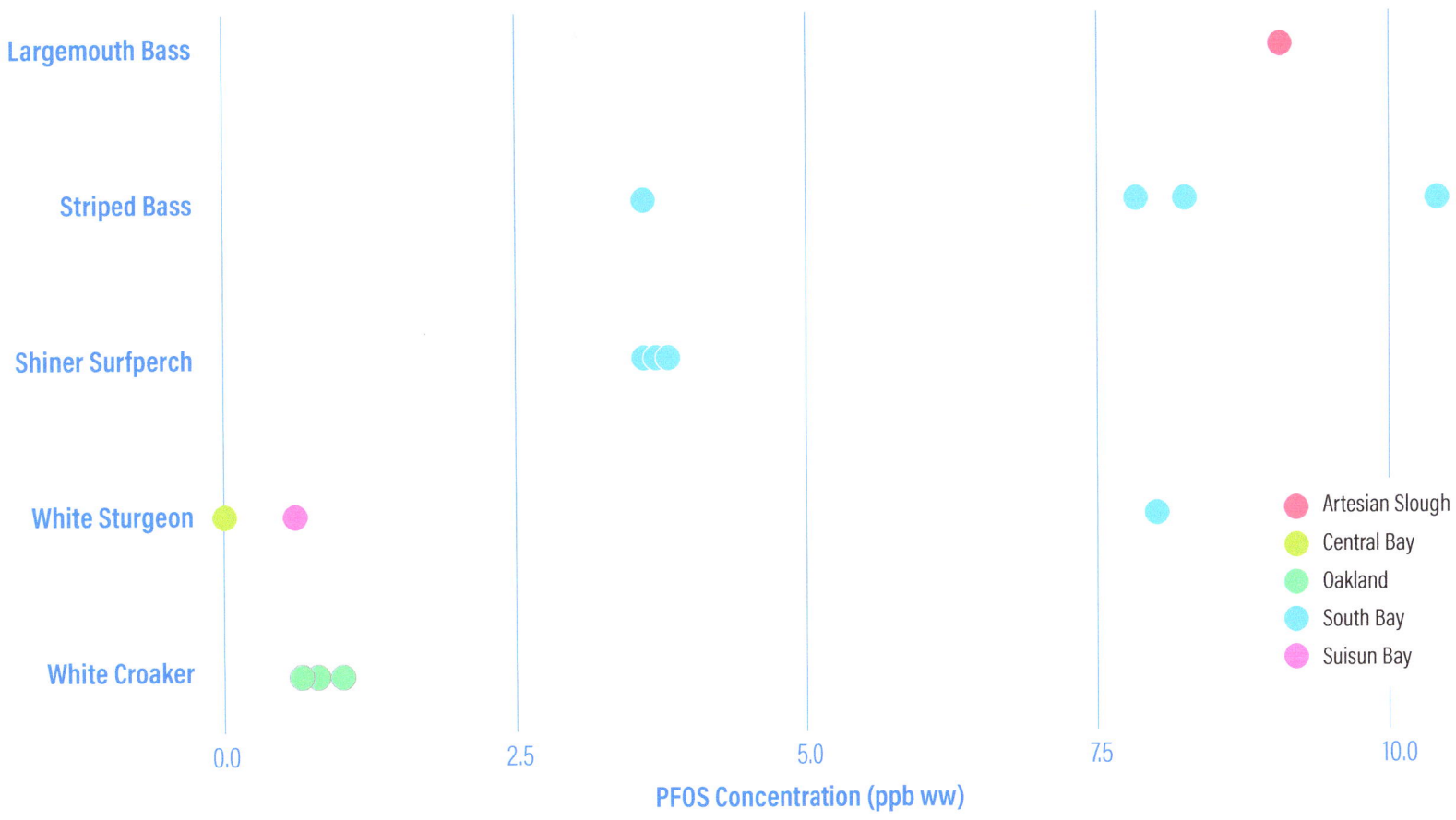

FOOTNOTE: PFOS concentrations (ppb ww) in San Francisco Bay fish, 2019. Bars indicate average concentrations. Points represent individual or composite samples. Points are colored by sampling location.

BISPHENOLS and OPEs

Origins of the Problem

Commercialization of plastics began in the 1950s, and has proceeded at such a rapid pace that by 2015, total global production since 1950 reached an estimated 7.8 billion metric tons — more than one ton of plastic for every person alive today. Plastics can be composed of a myriad of ingredients, from the chemicals that form the backbones of these synthetic polymers, to additives such as flame retardants. Two major classes of plastic ingredients are bisphenols and organophosphate esters (OPEs).

Bisphenol A (BPA), the most widely used and well-studied bisphenol, has been used to create polycarbonate plastics since the 1950s. Other uses include making epoxy resins such as those that line food and beverage cans, and thermal reactants for paper receipts. In 2019, the global production of BPA exceeded 8 million metric tons.

Organophosphate esters are also widely used in plastics, as additive flame retardants, and for various other purposes. Use as flame retardants increased dramatically following restrictions on the use of PBDE flame retardants in the 2000s (page 66). Global production of organophosphate esters in 2015 was estimated at 680,000 metric tons.

History of Monitoring

Very few data on these classes of contaminants were available for the Bay until the RMP launched a series of screening studies, starting in 2014 for organophosphate esters and 2017 for bisphenols. Observations across multiple matrices indicated concentrations of individual compounds in each class were approaching or exceeding protective thresholds for toxicity to aquatic life. These contaminants have also been observed at levels of interest in treated municipal wastewater and urban stormwater runoff discharging to the Bay.

History of Advisories and Regulation

Human health concerns led to bans of BPA in key products, with several states, including California, and the federal government implementing targeted restrictions since 2009. These bans have applied to a small number of products with food contact including baby bottles, sippy cups, and sports bottles; overall use of BPA remains high. BPA is an endocrine disruptor and is currently listed on California's Proposition 65 List for developmental and female reproductive toxicity, including a warning at the point of sale to address exposures via food packaging materials.

As the first BPA bans went into effect, some manufacturers began to use BPA alternatives including bisphenol F and S. These alternatives are not as well-studied as BPA, though available data suggest they share similar toxic properties. This can be considered a "regrettable substitution," the replacement of a toxic chemical with another, typically less-studied compound, which turns out to be just as harmful or even worse.

Organophosphate esters are recognized as regrettable substitutes for PBDEs, flame retardants that are now restricted in the US and many other countries (page 66). Some organophosphate esters are included on California's Proposition 65 List for carcinogenicity. In 2013, California's flammability standard for upholstered furniture, the driving force for the use of chemical flame retardants in these products, was updated to provide fire protection without the need for these additives. In 2017, the California Department of Toxic Substances Control's Safer Consumer Products Program established regulations on children's foam-padded sleeping products that led manufacturers to eliminate the use of a few especially-concerning organophosphate esters in their products. A subsequent 2018 state ban on all flame retardants in upholstered furniture, mattresses, and children's products (AB 2998), effective in 2020, is expected to limit the presence of organophosphate esters in these products. However, organophosphate esters are still used in many other products not covered by these management actions.

Current Status and Long-term Outlook

According to the United Nations Environment Programme, annual production of plastics is projected to increase from 335 million metric tons in 2016 to approximately 1124 million metric tons by 2050. The rapid growth of the plastics economy continues to drive demand for plastic ingredients such as bisphenols and organophosphate esters. The RMP currently considers both classes of contaminants to be moderate concerns for the Bay, and has plans for regular surveillance moving forward.

Observations by the RMP and the broader scientific community emphasize the need to control releases of these contaminants to protect water quality. The uses of these compounds are diverse; identifying major sources in urban environments, and assessing their potential to enter wastewater and stormwater pathways, is a critical next step to guide management actions via informed, rather than regrettable, substitution. §

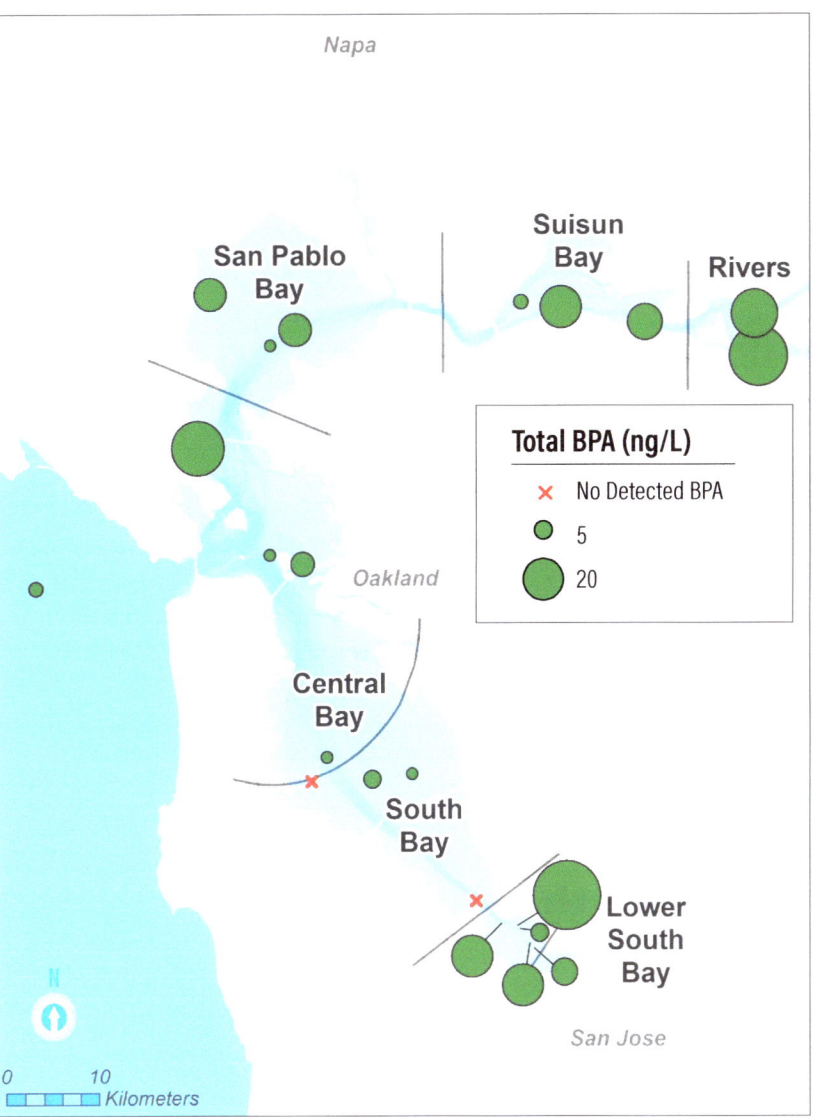

WATER QUALITY PARAMETER SUMMARIES **71**

MICROPLASTICS

Origins of the Problem

Plastics are among the most ubiquitous materials used in modern society. Microplastics, pieces of plastic under 5 mm in size, were first observed as contaminants in the oceans in 2004, and have since been identified in virtually every environment on Earth. Microplastics are often derived from larger plastic items, such as tiny tire wear particles shed while driving, fibers shed from textiles during washing and drying, and fragments from litter. Tire particles may be the biggest global source of microplastics. Due to our car culture, scientists estimate that the US has the highest tire particle emissions in the world—7 to 12 pounds per person every year.

History of Monitoring

In 2015, the RMP completed the first pilot study of microplastics in San Francisco Bay. This was closely followed by a first of its kind, comprehensive regional study. The San Francisco Bay Microplastics Project was completed in 2019, and found microplastics to be ubiquitous in Bay water, sediment, bivalves, and prey fish. This study quantified for the first time microplastics in urban stormwater runoff, and made the breakthrough discovery that concentrations in urban runoff were significantly higher than wastewater effluent. The vast majority of particles observed in urban stormwater runoff were suspected to be tire wear particles and fibers. Additionally in 2020, a collaboration with University of Washington identified various tire ingredients and tire-ingredient breakdown products in Bay stormwater runoff, including 6PPD-quinone at concentrations that are lethal to a salmon species that was historically present in the Bay (coho). More recent data indicate that steelhead, a salmon species still migrating through the Bay to surrounding watersheds, are also sensitive to this chemical.

Current Policy and Regulation

Early and ongoing policies related to microplastics have addressed discrete sources such as microbeads—bits of plastic used as ingredients in personal care products—and larger, commonly-littered plastic items like plastic bags and packaging. In 2018, California policy makers tasked state agencies to take leadership on microplastic and address the growing concerns about microplastic impacts on human and ecological health by passing and signing into law California Senate Bill 1263 (Portantino), which tasked the Ocean Protection Council with leading statewide efforts to address microplastic pollution by adopting and implementing a Statewide Microplastics Strategy. The Strategy, presented to the California legislature in 2022, acknowledges the importance of tire wear particles and calls for the development of a tires-specific pollution prevention strategy.

In 2022, the California Department of Toxic Substances Control proposed regulations that would require tire manufacturers to identify alternatives to the tire ingedient 6PPD, which readily transforms into 6PPD-quinone in the environment.

Current Status and Long-term Outlook

Microplastics are prevalent throughout the Bay. These contaminants are persistent, because plastics fragment into smaller and smaller pieces in the environment and take decades or centuries to fully degrade. As a result, microplastic concentrations will build up in the Bay over time if society continues with business-as-usual increases in plastic use.

DTSC's regulatory efforts to address toxic tire ingredients like 6PPD may lead industry to adopt safer alternatives, although market changes can take many years. Meanwhile, no actions have yet been implemented to address fibers, the most common type of microplastic observed in the Bay and consumed by wildlife. Levels of tire wear particles, fibers and other microplastics will rise without clear policy goals and management actions. The RMP is making strategic investments to implement science that will guide these management actions. §

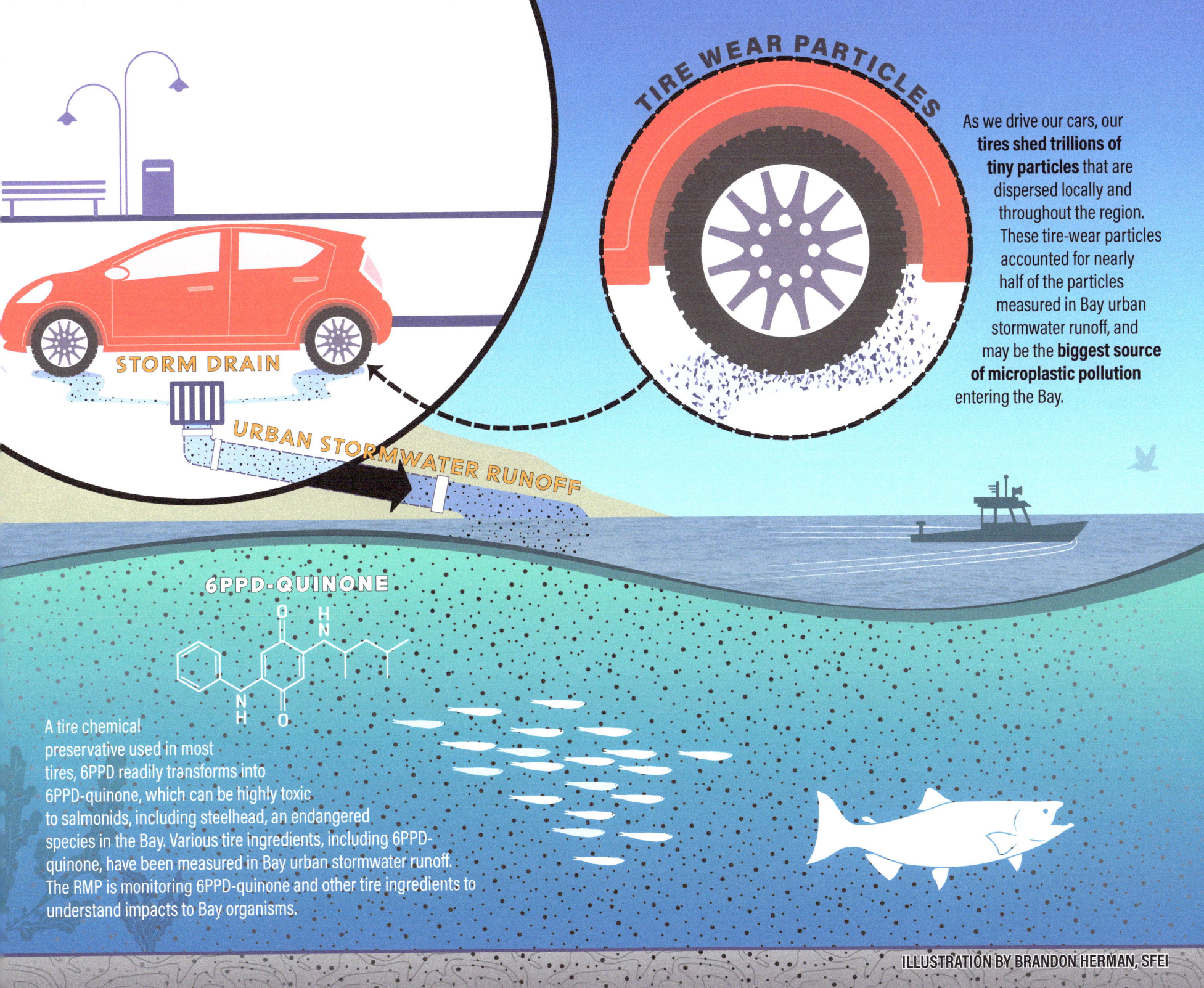

303(d) LIST

Section 303(d) of the 1972 Federal Clean Water Act requires that states develop a list of water bodies that do not meet water quality standards and develop action plans, called Total Maximum Daily Loads (TMDLs), to improve water quality.

The list of impaired water bodies is periodically updated. The RMP is one of many entities that provide data to the State Water Board to assess water quality and inform the 303(d) List. The process for developing the 303(d) List for the Bay includes the following steps:

- preparation of a draft list of listing/delisting recommendations by Regional Water Board staff;
- adoption by the State Water Board; and
- approval by USEPA.

The primary pollutants/stressors for the Bay and its major tributaries on the 303(d) List include:

Trace elements: Mercury and Selenium

Pesticides: Dieldrin, Chlordane, and DDT

Other chlorinated compounds:
PCBs
Dioxin and Furan Compounds

Others: Exotic Species, Trash, Polycyclic Aromatic Hydrocarbons (PAHs), and Indicator Bacteria

STATUS OF POLLUTANTS OF CONCERN

PARAMETER	STATUS
Copper	Site-specific objectives approved for entire Bay
	San Francisco Bay removed from 303(d) List in 2002
Dioxins / Furans	Monitoring recovery (synthesis report prepared in 2018)
Legacy Pesticides (Chlordane, Dieldrin, and DDT)	Monitoring recovery
Mercury	Bay TMDL and site-specific objectives approved in 2008
	Guadalupe River Watershed TMDL approved in 2010
Bacteria	Richardson Bay TMDL approved in 2008
	Bay beaches (multiple listings); TMDL approved in 2017
	Bay beaches (additional multiple listings); second TMDL in development
PCBs	Bay TMDL approved in 2010
Selenium	North Bay TMDL approved in 2016
Trash	Municipalities required to implement trash load controls in 2009
Dissolved Oxygen	Site-specific objectives for Suisun Marsh approved in 2019

RMP COMMITTEE MEMBERS AND PARTICIPANTS

RMP Steering Committee
BACWA Principal, Eric Dunlavey, City of San Jose
BACWA Associate, Amanda Roa, Delta Diablo
BACWA Associate, Karin North, City of Palo Alto
Stormwater Agencies, Adam Olivieri, EOA, Inc.
Dredgers, John Coleman, Bay Planning Coalition
Industrial Wastewater, Maureen Dunn, Chevron
San Francisco Bay Regional Water Quality Control Board, **Tom Mumley**
US Army Corps of Engineers, Tessa Beach
RMP Steering Committee Chair in bold

RMP Technical Review Committee
BACWA, Mary Lou Esparza, Central Contra Costa Sanitary District
BACWA, Yuyun Shang, East Bay Municipal Utility District
South Bay Dischargers, Tom Hall, EOA, Inc.
City and County of San Francisco, Heather Peterson
City of San Jose, Anne Hansen Balis
Refineries, **Bridgette DeShields**, Integral Consulting Inc.
Stormwater, Chris Sommers, EOA, Inc.
Dredgers, Shannon Alford, Port of San Francisco
San Francisco Bay Regional Water Quality Control Board, Richard Looker
USEPA Region IX, Luisa Valiela
US Army Corps of Engineers, Tessa Beach
NGO, Ian Wren, Baykeeper
RMP Technical Review Committee Chair in bold

RMP Science Advisors

EMERGING CONTAMINANTS WORKGROUP
Dr. Bill Arnold, University of Minnesota
Dr. Lee Ferguson, Duke University
Dr. Derek Muir, Environment Canada
Dr. Heather Stapleton, Duke University
Dr. Miriam Diamond, University of Toronto
Dr. Dan Villeneuve, USEPA

MICROPLASTICS WORKGROUP
Dr. Chelsea Rochman, University of Toronto

PCB WORKGROUP
Dr. Frank Gobas, Simon Fraser University
Dr. Earl Hayter, US Army Engineer Research and Development Center

SEDIMENT WORKGROUP
Dr. David Schoellhamer, USGS Emeritus
Dr. Patricia Wiberg, University of Virginia

SOURCES, PATHWAYS, AND LOADINGS WORKGROUP
Dr. Robert Budd, California Department of Pesticide Regulation
Dr. Jon Butcher, Tetra Tech
Steve Corsi, USGS
Tom Jobes, Independent

RMP Participants

MUNICIPAL DISCHARGERS
City of American Canyon
City of Benicia
City of Burlingame
City of Calistoga
Central Contra Costa Sanitary District
Central Marin Sanitation Agency
Delta Diablo
East Bay Dischargers Authority
East Bay Municipal Utility District
Fairfield-Suisun Sewer District
Las Gallinas Valley Sanitary District
City of Millbrae
Mountain View Sanitary District
Napa Sanitation District
Novato Sanitation District
City of Palo Alto
City of Petaluma
City of Pinole/Hercules
Rodeo Sanitary District
San Francisco International Airport
City and County of San Francisco
City of San Jose
City of San Mateo
Sausalito-Marin City Sanitary District
Sewerage Agency of Southern Marin
City of South San Francisco/San Bruno
Sonoma County Water Agency
Silicon Valley Clean Water
City of Sunnyvale
City of St. Helena
Marin County Sanitary District #5, Tiburon
Union Sanitary District
Vallejo Flood and Wastewater District
West County Wastewater District
Town of Yountville
U.S. Navy, Treasure Island

INDUSTRIAL DISCHARGERS
C&H Sugar Company
Chevron Products Company
Crockett Cogeneration
Eco Services Operations Corporation
Marathon Petroleum
PBF Martinez Refining Company
Phillips 66 Company
Schnitzer Steel Industries
USS-POSCO Industries
Valero Refining Company

STORMWATER
Alameda County Clean Water Program
California Department of Transportation
City and County of San Francisco
Contra Costa Clean Water Program
Fairfield-Suisun Urban Runoff Management Program
Marin County Stormwater Pollution Prevention Program
Santa Clara Valley Urban Runoff Pollution Prevention Program
San Mateo Countywide Water Pollution Prevention Program
Vallejo Sanitation & Flood Control District

DREDGERS
Chevron Richmond Long Wharf
City of Benicia Marina
City of Petaluma
Contra Costa Water District
Kinder Morgan Richmond Terminal
Loch Lomond Marina
Oyster Cove Marina
Paradise Caye Yacht Harbor
Phillips 66 Company Rodeo Terminal
Port of Oakland
Port of Redwood City
Port of San Francisco
Richardson Bay Marina
San Francisco Yacht Club
Valero Refinery Terminal
Westpoint Harbor
Vallejo Yacht Club

Tepco Beach · Shira Bezalel, August 2022

REFERENCES

Section 1

Dow, G.R. 1973. Bay Fill in San Francisco: A History of Change. Master's Thesis, San Francisco State University, San Francisco, CA. https://semspub.epa.gov/work/09/1137835.pdf

Hyde, G.C., H.F Gray, A.M. Rawn. 1941. East Bay Cities Sewage Disposal Survey, Report Upon the Collection, Treatment and Disposal of Sewage and Industrial Wastes of the East Bay Cities, California. The Board of Consulting Engineers. https://babel.hathitrust.org/cgi/pt?id=mdp.39015006094711&view=1up&seq=5

Newton, C.R. 1990. Dredging in the United States. Theses and Major Papers. Paper 202. https://digitalcommons.uri.edu/ma_etds/202/

USACE. 1975. Final Composite Environmental Statement for Maintenance Dredging of Existing Navigation Projects in San Francisco Bay (1975). US Army Corps of Engineers, San Francisco, CA. https://www.spn.usace.army.mil/Missions/Dredging-Work-Permits/LTMS/December-1975-Volume-1/

USACE (US Army Corps of Engineers), US Environmental Protection Agency, San Francisco Bay Conservation and Development Commission, and San Francisco Bay Regional Water Quality Control Board. 1998. Long-Term Management Strategy for the Placement of Dredged Material in the San Francisco Bay Region. Policy Environmental Impact Statement/Programmatic Environmental Report. https://www.spn.usace.army.mil/Missions/Dredging-Work-Permits/LTMS/

USACE (US Army Corps of Engineers), US Environmental Protection Agency, San Francisco Bay Conservation and Development Commission, and San Francisco Bay Regional Water Quality Control Board. 2001. Long-Term Management Strategy for the Placement of Dredged Material in the San Francisco Bay Region, Management Plan. 2001. https://www.spn.usace.army.mil/Portals/68/docs/Dredging/LMTS/entire%20LMTF.pdf

USEPA/USACE. 1991. Evaluation of Dredged Material Proposed for Ocean Disposal - Testing Manual. Report No. EPA 583/8-91/001. Office of Water. https://www.epa.gov/sites/default/files/2015-10/documents/green_book.pdf

USEPA/USACE. 1998. Evaluation of Dredged Material Proposed for Discharge in Waters of the U. S. - Testing Manual. Report No. EPA 823-B-94-002. Office of Water. https://www.epa.gov/sites/default/files/2015-08/documents/inland_testing_manual_0.pdf

Section 2

Beck, M.W., de Valpine, P., Murphy, R., Wren, I., Chelsky, A., Foley, M. and Senn, D.B., 2022. Multi-scale trend analysis of water quality using error propagation of generalized additive models. Science of the Total Environment, 802, p.149927.

Cloern, J.E. and Jassby, A.D., 2012. Drivers of change in estuarine-coastal ecosystems: Discoveries from four decades of study in San Francisco Bay. Reviews of Geophysics, 50(4).

Cloern, J.E., Jassby, A.D., Thompson, J.K. and Hieb, K.A., 2007. A cold phase of the East Pacific triggers new phytoplankton blooms in San Francisco Bay. Proceedings of the National Academy of Sciences, 104(47), pp.18561-18565.

Cloern, J.E., Schraga, T.S., Nejad, E. and Martin, C., 2020. Nutrient status of San Francisco Bay and its management implications. Estuaries and Coasts, 43(6), pp.1299-1317.

Heal the Bay. 2022. 2021-22 Beach Report Card. Heal the Bay, Santa Monica, CA

Peacock, M.B., Gibble, C.M., Senn, D.B., Cloern, J.E. and Kudela, R.M., 2018. Blurred lines: Multiple freshwater and marine algal toxins at the land-sea interface of San Francisco Bay, California. Harmful Algae, 73, pp.138-147.

Roberts, D., MacVean, L., Holleman, R., Chelsky, A., Art, K., Nidzieko, N., Sylvester, Z. and Senn, D., 2022. Connections to Tidal Marsh and Restored Salt Ponds Drive Seasonal and Spatial Variability in Ecosystem Metabolic Rates in Lower South San Francisco Bay. Estuaries and Coasts, pp.1-18.

Schraga, T.S. and Cloern, J.E., 2017. Water quality measurements in San Francisco Bay by the US Geological Survey, 1969–2015. Scientific Data, 4(1), pp.1-14.

SFBRWQCB. 2000. 50 Years of Protecting Bay Area Waters. San Francisco Bay Regional Water Quality Control Board, Oakland, CA.

SFEI. 2014a. Scientific Foundation for the San Francisco Bay Nutrient Management Strategy, San Francisco Estuary Institute, Contribution #979. link

SFEI. 2014b. External Nutrient Loads to San Francisco Bay. San Francisco Estuary Institute, Contribution #704. link

SFEI. 2015. Lower South Bay Nutrient Synthesis. San Francisco Estuary Institute, Contribution #732 link

SFEI. 2018. Dissolved Oxygen in Lower South Bay: Variability, Important Processes, and Implications for Understanding Fish Habitat. San Francisco Estuary Institute, Contribution #911. link

SFEI. In prep. Trends in nutrient-related water quality indicators in San Francisco Bay: 1990-2019.

Sutula, M., Kudela, R., Hagy III, J.D., Harding Jr, L.W., Senn, D., Cloern, J.E., Bricker, S., Berg, G.M. and Beck, M., 2017. Novel analyses of long-term data provide a scientific basis for chlorophyll-a thresholds in San Francisco Bay. Estuarine, Coastal and Shelf Science, 197, pp.107-118.

CREDITS AND ACKNOWLEDGEMENTS

EDITORS
Jay Davis
Melissa Foley

DESIGN and PRODUCTION
Ruth Askevold and Ellen Plane

CONTRIBUTING AUTHORS
Jay Davis
Rebecca Sutton
Melissa Foley
Dave Senn
Diana Lin
Miguel Mendez

RMP DATA MANAGEMENT and QUALITY ASSURANCE
Cristina Grosso
Don Yee
Adam Wong
Michael Weaver

INFORMATION COMPILATION
Richard Looker
Adam Wong

INFORMATION GRAPHICS
Brandon Herman

HISTORICAL PHOTO PROJECT
Photography: Shira Bezalel (www.shirabezalelphotography.com/)

Research: Ruth Askevold, Shira Bezalel, Sean Baumgarten, Jay Davis

Historical Image Permissions: Alex Cherian (Bay Area TV Archive, San Francisco State University), Fabiola Franco (Nexstar Broadcasting Company)

Information and Images: Eric Dunlavey (City of San Jose), Timothy Grillo (Union Sanitary District), Lorien Fono (BACWA)

Historical San Francisco-Oakland Bay Bridge image: Courtesy of Library of Congress

Historical Golden Gate Bridge image: Courtesy of photovault.com

REVIEW
The following reviewers greatly improved this document by providing comments on draft versions:

Tom Hall
Richard Looker
Kelly Moran
Don Yee
Lester McKee
Andy Gunther
Ian Wren
Luisa Valiela
Adam Olivieri
Bridgette DeShields
Barbara Baginska
Paul Amato
Jim Haussener

Black-crowned Night-heron at Bay's edge · Shira Bezalel

REGIONAL MONITORING PROGRAM FOR WATER QUALITY IN SAN FRANCISCO BAY
sfei.org/rmp

Administered by the San Francisco Estuary Institute
4911 Central Avenue, Richmond, CA 94804
p: 510-746-SFEI (7334), f: 510-746-7300

www.sfei.org

www.ingramcontent.com/pod-product-compliance
Lightning Source LLC
Chambersburg PA
CBHW042036120526
44592CB00029B/85